THE BIOLOGY OF
PLASMIDS

The Biology of Plasmids

DAVID K. SUMMERS
MA, DPhil
Department of Genetics, University of Cambridge,
Downing Street, Cambridge CB2 3EH

Blackwell
Science

© 1996 by
Blackwell Science Ltd
Editorial Offices:
Osney Mead, Oxford OX2 0EL
25 John Street, London WC1N 2BL
23 Ainslie Place, Edinburgh EH3 6AJ
238 Main Street, Cambridge
  Massachusetts 02142, USA
54 University Street, Carlton
  Victoria 3053, Australia

Other Editorial Offices:
Arnette Blackwell SA
  224, Boulevard Saint Germain,
  75007 Paris, France

Blackwell Wissenschafts-Verlag GmbH
  Kurfürstendamm 57
  10707 Berlin, Germany

  Zehetnergasse 6, A-1140 Wien
  Austria

First published 1996

Set by Semantic Graphics, Singapore

DISTRIBUTORS

Marston Book Services Ltd
PO Box 87
Oxford OX2 0DT
(*Orders:* Tel: 01865 791155
       Fax: 01865 791927
       Telex: 837515)

USA
Blackwell Science, Inc.
238 Main Street
Cambridge, MA 02142
(*Orders:* Tel: 800 215-1000
       617 876-7000
       Fax: 617 492-5263)

Canada
Copp Clark, Ltd
2775 Matheson Blvd East
Mississauga, Ontario
Canada, L4W 4P7
(*Orders*: Tel: 800 263-4374
       905 238-6074

Australia
Blackwell Science Pty Ltd
54 University Street
Carlton, Victoria 3053
(*Orders:* Tel: 03 9347 0300
       Fax: 03 9349 3016)

A catalogue record for this title
is available from the British Library

ISBN 0-632-03436-X

Library of Congress
Cataloging-in-Publication Data

Summers, David K.
  The biology of plasmids /
  David K. Summers.
       p.   cm.
  Includes   bibliographical
  references
  and index.
  ISBN 0-632-03436-X
  1. Plasmids.
  2. Plasmids—Genetics.
  I. Title.
QR76.6.S85   1996
589.9'08732—dc20   95-42723
                             CIP

Contents

Preface

As a student in search of a research project in the late 1970s, plasmid biology seemed to be one of very few routes by which I could enter the new and exciting field of molecular biology. Like many bacteria before me I found that, once picked up, plasmids were hard to put down. Fifteen years later they continue to provide excellent model systems with which to study the most fundamental life processes at the molecular level.

This book is about the biology of plasmids. Their use and abuse has been described exhaustively elsewhere and is not repeated here. Instead I have chosen to put on paper the arguments I use when trying to persuade sceptical undergraduates that plasmids deserve their recognition and respect as some of the simplest and most highly evolved living organisms.

As a geneticist I can hardly avoid thanking my parents for making this book possible and perhaps, more significantly, my children for making it necessary. Intellectual credits (but none of the blame) go to Philip Oliver who introduced me to bacterial plasmids as an undergraduate, to Keith Dyke who saw me through my apprenticeship in Oxford and to David Sherratt for providing an environment at Glasgow University in the early 1980s where plasmid biology was at once exciting and entertaining. Most of all I owe my gratitude to my wife who during the writing of this book has had to tolerate emotional swings between elation and despair the like of which we have not seen since the writing of my DPhil thesis.

David Summers
Cambridge

1: The Anatomy of Bacterial Plasmids

Prologue

The chromosome is a repository of indispensable genes encoding the housekeeping functions of bacteria. Here we can find that potential for growth and division which, if unchecked, would allow a single *Escherichia coli* to produce more progeny between breakfast and tea-time than the total human population of the planet. Here we find genes which have remained virtually unaltered over unimaginable periods of time; the meat and potatoes of prokaryotic cuisine. For variety and diversity, however, for champagne and caviar we must look to plasmids. These cytoplasmic gene clusters borrow host functions for the majority of their DNA metabolism yet remain physically separate from the chromosome. It is by appreciating the genetic independence of these elements that we can understand how they have played a central role in bacterial evolution, and in the emergence of molecular biology as the most exciting area of science in the late twentieth century.

The first word

What are the origins of plasmid biology? The name itself is relatively recent, first appearing in print in the *Physiological Reviews* of 1952.

> These discussions have left a plethora of terms adrift: pangenes, plastogenes, chondriogenes, cytogenes and proviruses, which have lost their original utility owing to the accretion of vague or contradictory connotations. At the risk of adding to this list, I propose plasmid as a generic term for any extrachromosomal hereditary determinant. The plasmid may be genetically simple or complex. *(Joshua Lederberg (1952))*

Early interest in plasmids arose from two very different quarters. Of interest originally only to microbial geneticists as the agents of pseudosexual gene exchange, they were brought to the attention of a much wider public when they were found to be responsible for the spread of antibiotic resistance.

1.1 Early plasmid research

1.1.1 Fertile beginnings

During the 1940s and early 1950s studies of recombination in *E. coli* K12 revealed that the transfer of genetic information in bacterial matings is unidirectional. Sexual differentiation was attributed to possession of a transmissible factor called F (for fertility) with F^+ cells acting as donors of information and F^- cells as recipients. The physical nature of F was an enigma in those early years; it was described variously as 'an ambulatory or infective hereditary factor' (Lederberg *et al.*, 1952) and 'a non-lytic infectious agent' (Hayes, 1953); the latter drawing parallels with the established ability of bacteriophage to mediate gene transfer.

Almost a decade after the discovery of bacterial conjugation came a rigorous demonstration that the F factor was made of DNA, when it was isolated on a caesium chloride density gradient (Marmur *et al.*, 1961). In order to separate plasmid and chromosomal DNA Marmur ingeniously transferred the F factor to *Serratia marcescens*, whose chromosome contains 58% G + C compared to 50% G + C for F and the *E. coli* chromosome. The presence of F in the new host correlated with the appearance of a new band on the density gradient at a position corresponding to 50% G + C, in addition to the *Serratia* chromosomal band at 58%.

1.1.2 Plasmids and antibiotic resistance

In early studies, interest in F stemmed purely from its utility as a tool to facilitate genetic analysis in *E. coli*. The situation changed dramatically, however, in the 1950s and 1960s when plasmids were found to be responsible for the spread of multiple antibiotic resistance. After the Second World War the use of antibacterial drugs to combat infection became increasingly common. From the earliest days, strains of bacteria resistant to drugs such as streptomycin had been isolated in the laboratory using conventional techniques of mutation and selection. Resistant strains arose at low frequencies and the mutations responsible mapped to loci on the bacterial chromosome. Typically, they conferred resistance by altering the target of the drug. The relative ease with which resistant strains of bacteria could be isolated in the laboratory meant that it came as little surprise when clinical isolates showed resistance to single antibiotics, but the appearance and rapid spread of multiply resistant strains (considered in detail in chapter 5) was as unexpected as it was unwelcome. One unexpected benefit, however, was the transformation of plasmid biology from an academic curiosity into one of the most exciting and rapidly expanding areas of microbial genetics.

1.2 Plasmid-encoded phenotypes

1.2.1 The diversity of plasmid-borne genes

In addition to antibiotic resistance and bacterial conjugation, plasmids encode an enormous variety of functions which are not essential for survival of the host or plasmid but which extend host range by increasing fitness in atypical environments (Table 1.1). One class of plasmids protects the host from the ill effects of heavy metals, toxic anions and intercalating agents and, through provision of additional repair systems, confers increased resistance to ionizing radiations. A second class extends the host's metabolic versatility. This group includes plasmids which encode enzymes for the synthesis of bioactive compounds including colicins and antibiotics, or which confer the ability to degrade recalcitrant organic molecules. Thirdly we find plasmids which open up new environments for their bacterial host, conferring pathogenicity by encoding toxins and colonization antigens.

1.2.2 Bacteriocin production and resistance

Among the most bizarre of plasmid-encoded traits is the production of bactericidal agents coupled with immunity from their effects. In 1925 the Belgian microbiologist André Gratia observed that certain strains of bacteria secrete proteins that kill non-producing strains. These are known generally as bacteriocins or colicins when produced by coliform bacteria. Plasmids which encode bacteriocins are widespread; in a collection of 433 bacterial strains made between 1917 and 1954, 59 were colicinogenic (Datta, 1985). The genetic determinants for colicin production and resistance have been studied in detail for two groups of Col plasmids. Group I consists of small (3–6 MDa) high copy number plasmids such as ColE1, while group II contains large (70–120 MDa) low copy plasmids including ColIB. Colicins and the Col plasmids which encode them have been reviewed by Luria and Suit (1987).

1.2.3 Colicin production

A well-studied example of colicin production and its consequences is provided by the multicopy plasmid ColE1. Production of the toxin requires the simultaneous expression of the colicin structural gene (*cea*) and a second gene (*kil*), which is responsible for cell lysis

Table 1.1 Plasmid phenotypes. (After Stanish, *Methods Microbiol.* 2(1)).

1 Resistance properties

Antibiotic resistance
Aminoglycosides (e.g. streptomycin, gentamicin, amikacin)
Chloramphenicol
Fusidic acid
β-Lactam antibiotics (e.g. benzyl penicillin, ampicillin, carbenicillin)
Sulphonamides, trimethoprim
Tetracyclines
Macrolides (e.g. erythromycin)

Heavy metal resistance
Mercuric ions and organomercurials
Nickel, cobalt, lead, cadmium bismuth, antimony, zinc, silver

Resistance to toxic anions
Arsenate, arsenite, tellurite, borate, chromate

Other resistances
Intercalating agents (e.g. acridines, ethidium bromide)
Radiation damage (e.g. by UV light, X-rays)
Bacteriophage and bacteriocins
Plasmid-specified restriction/modification systems

2 Metabolic properties
Antibiotic production
Bacteriocin production
Metabolism of simple carbohydrates (e.g. lactose, sucrose, raffinose)
Metabolism of complex carbon compounds (e.g. octane, toluene, camphor,
  nicotine, aniline) and halogenated compounds (e.g. 2,6-dichlorotoluene,
    2,4-dichlorophenoxyacetic acid)
Metabolism of proteins (e.g. casein, gelatin)
Metabolism of opines (by Ti$^+$ *Agrobacterium*)
Nitrogen fixation (by Nif$^+$ *Rhizobium*)
Citrate utilization
Phosphoribulokinase activity in *Alcaligenes*
Thiamine synthesis by *Erwinia* and *Rhizobium*
Denitrification activity in *Alcaligenes*
Proline biosynthesis by Ti$^+$ *Agrobacterium*
Pigmentation in *Erwinia*
H$_2$S production
Extracellular DNase

3 Factors modifying host lifestyle
Toxin production
Enterotoxins of *Escherichia coli*
Exfoliative toxin of *Staphylococcus aureus*
Exotoxin of *Bacillus anthracis*
δ-endotoxin of *Bacillus thuringiensis*
Neurotoxin of *Clostridium tetani*

Colonization antigens of *Escherichia coli* (e.g. K88, K99, CFAI, CFAII)
Haemolysin synthesis (e.g. in *Escherichia coli* and *Streptococcus*)
Serum resistance of enterobacteria
Virulence of *Yersinia* species

Continued

Table 1.1 (*Continued*)

3. Factors modifying host lifestyle (*Continued*)
Capsule production of *Bacillus anthracis*
Crown gall and hairy root disease of plants (by Ti+ and Ri+ *Agrobacterium*)
Infection and nodulation of legumes (by Sym+ *Rhizobium*)
Iron transport (e.g. in *Escherichia coli* and *Vibrio anguillarum*)

4 Miscellaneous properties
Gas vacuole formation in *Halobacterium*
Pock formation (lethal zygosis) in *Streptomyces*
Killing of *Klebsiella pneumoniae* by Kik+ IncN plasmids
Sensitivity to bacteriocins in *Agrobacterium*
Translucent/opaque colony variation in *Mycobacterium*
Rhizosphere protein by Nod+ Fix+ *Rhizobium leguminosarum*
R-inclusion body production in *Caedibacter*
Endopeptidase activity by *Staphylococcus*
Chemotaxis towards acetosyringone by Ti+ *Agrobacterium*

(Fig. 1.1). In most cells transcription of *cea* and *kil* is repressed by the LexA protein but when the SOS response is triggered by DNA damage, the repressor is inactivated by RecA protease activity. Only one in 10^{-3}–10^{-4} cells produces colicin each generation and this suicidal act poisons the surrounding environment. Death of the colicin producer is apparently the result of *kil*-mediated lysis rather than colicin synthesis because the induction of colicin production from a *kil*⁻ *imm*⁺ *cea*⁺ plasmid is not lethal. The level of colicin expression after exA repression is lifted, is modulated by the host's metabolic state and is some 45-fold greater in anaerobic than in aerobic conditions (Eraso and Weinstock, 1992).

Protection against exogenous colicin is conferred by the plasmid immunity (*imm*) gene. The occasional induction of *cea* and *kil* maintains a low concentration of colicin in the immediate environment of bacteria carrying Col plasmids, ensuring immunity from invasion by cells lacking the plasmid-borne immunity gene (Fig. 1.2). In many respects, the consequences of the acquisition of a colicin plasmid or a lysogenic bacteriophage are strikingly similar. Induction of both colicin expression and the phage lytic cycle are linked to the

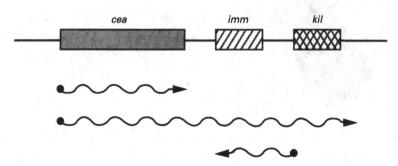

Fig. 1.1 The colicin cassette of plasmid ColE1 contains the colicin structural gene *cea*, the immunity gene *imm* and the gene responsible for cell lysis, *kil*. Arrows indicate the extent and direction of transcripts.

SOS response, both kill the host cell and both result in the release of a bactericidal agent (colicin or phage particles) into the environment. Cultures carring either ColE1 or a λ lysogen are equally protected from competition by invaders lacking the respective plasmid or phage immunity genes.

Col plasmids provide bacteria with the means to manipulate their environment and maximize fitness in competitive situations. However, the initial proportion of colicin-producers in a new niche is inevitably low. Chao and Levin (1981) have shown that in liquid culture the colicin plasmid must be carried by at least 2% of cells before any advantage is gained. Below this level the exogenous colicin concentration is so low that the benefits associated with the inhibition of sensitive competitors is less than the resources expended to fulfil the metabolic demands of the Col plasmid. If L-broth was the natural environment of *E. coli*, it would be hard to understand how Col plasmids could help their hosts to repel invaders or invade new niches (Levin, 1988). In a soft agar matrix or other structured habitat, however, the situation is quite different. Bacteria grow as discrete colonies and, by release of colicin into their surroundings, establish a local exclusion zone in which sensitive neighbours are killed. Under these conditions, colicin producers gain an advantage at frequencies of less than 10^{-6} and it is likely that structured habitats (such as the mammalian large intestine) have played a crucial role in the evolution of colicin and the proliferation of colicin-producing strains.

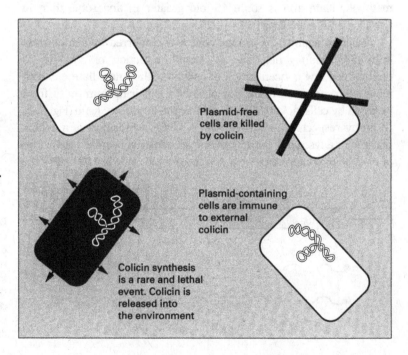

Fig. 1.2 Rare induction of the ColE1 *cea* and *kil* genes leads to the release of colicin, rendering the environment toxic for plasmid-free cells. Although synthesis and release of colicin is invariably lethal, plasmid-containing cells are immune to external colicin.

Plasmid-free cells are killed by colicin

Plasmid-containing cells are immune to external colicin

Colicin synthesis is a rare and lethal event. Colicin is released into the environment

1.3 Plasmid classification

1.3.1 Early attempts

As more and more plasmids were discovered and the bewildering diversity of plasmid-encoded functions was catalogued, it became essential to develop a practical system of classification. Early attempts were based on phenotypes such as antibiotic resistance and bacteriocin production but a desire to use some more fundamental property led to a system of classification based upon transmissibility. R factors were classified as fi^+ (fertility inhibition-plus) or fi^- according to their ability to suppress transfer of plasmid F. Within each group, plasmids were further classified by a variety of criteria including pilus type, base ratios and susceptibility to host-mediated restriction (Meynell *et al.*, 1968).

None of these early systems of classification was entirely satisfactory, particularly as an increasing number of non-conjugative plasmids were being discovered. There was a need for a scheme based upon a universal plasmid property and the most obvious was replication. In practice, replication control systems are compared by studying the incompatibility relationships between plasmids and by comparing DNA sequences of their basic replicons. Modern approaches to plasmid classification have been reviewed by Couturier *et al.* (1988).

1.3.2 Incompatibility classification

'Incompatibility' is defined as the inability of two plasmids to stably coexist in the same cell line. Thus a cell containing incompatible plasmids A and B will give rise eventually to descendants containing either A or B, but not both. Strong incompatibility results from a cross-reaction between replication control systems and a weaker effect is seen between plasmids which employ closely related partition functions. A formal system of classification based upon incompatibility was developed in the early 1970s with pairs of incompatible plasmids being assigned to the same incompatibility group. At present some 30 groups have been defined among plasmids of enteric bacteria and seven among the staphylococcal plasmids.

The practicalities of plasmid classification have been discussed by Bergquist (1987). The most common form of incompatibility test involves the introduction of plasmid A by transformation or conjugation into a strain in which plasmid B is already established. Selection is for plasmid A alone and transductant or transconjugant colonies are screened for the presence of plasmid B. If B is no longer present, the plasmids are pronounced incompatible. If all colonies tested contain both plasmids it is assumed that they are fully compatible. The result is not always clear-cut, however, and some transformants may contain only A while others retain both A and B. Further experiments may be necessary to clarify the result using the maintenance or segregation test. The maintenance test involves non-selective culturing of colonies containing both plasmids and screening for the loss of either. A slightly different approach is used in the segregation test where, after initial introduction of plasmid A, transformants are subcultured under conditions which select for plasmid A and loss of plasmid B is monitored.

1.3.3 The molecular basis of plasmid incompatibility

Plasmid incompatibility is due primarily to interference between replication control functions. Most plasmids produce a *trans*-acting repressor of replication whose concentration is proportional to plasmid copy number. The interaction of the repressor with its target breaks a link in the chain of events leading to the initiation of replication and establishes a negative feedback loop which regulates plasmid copy number. When copy number and repressor concentration are high, plasmid replication is inhibited but when copy number and repressor concentration are low, replication proceeds.

Consider a cell containing two compatible plasmids. Each plasmid produces a replication inhibitor which has no effect on the replication of the other; they maintain their normal copy numbers and persist independently of one another. In contrast, each member of a pair of incompatible plasmids produces an inhibitor which regulates not only its own replication but also that of its cell-mate (Fig. 1.3). A consequence of cross-inhibition is that the total number of plasmid copies in the cell is less than the sum of the normal individual copy numbers because each plasmid responds to the *total* inhibitor concentration, adjusting its replication rate and copy number accordingly. This alone, however, is insufficient to explain the segregation of incompatible plasmids. Segregation occurs because the copy number control systems fail to distinguish between the two plasmids which constitute a single pool from which representatives are selected at random for replication. Variation in the number of replications that each plasmid achieves *per* generation means that although the total copy number is regulated, the relative contribution from the two incompatible plasmids will drift and, in time, the loss of one or other is inevitable.

If the incompatible plasmids are of equal copy number and if each inhibitor efficiently represses replication of both, the incompatibility reaction will be symmetrical. Cell lines containing either plasmid will arise with equal probability. Sometimes, however, copy numbers are not the same; the higher copy number plasmid may be less sensitive to the replication inhibitor. This plasmid will continue to replicate at inhibitor concentrations which abolish replication of its low copy number counterpart. Being thus deprived of the ability to replicate, the low copy plasmid will be lost preferentially from mixed cells.

1.3.4 Shortcomings of incompatibility testing

The main virtue of incompatibility testing as a basis for classification is that it exploits a feature common to all plasmids. However, although it is generally accepted that incompatibility studies provide an indication of evolutionary relatedness, the results of the tests can sometimes be both confusing and misleading. Many plasmids

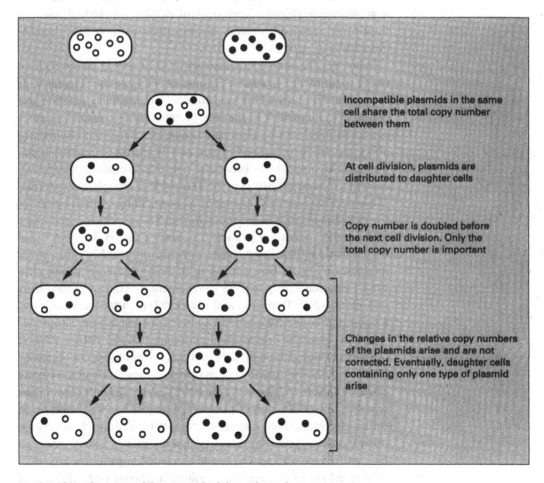

Incompatible plasmids in the same cell share the total copy number between them

At cell division, plasmids are distributed to daughter cells

Copy number is doubled before the next cell division. Only the total copy number is important

Changes in the relative copy numbers of the plasmids arise and are not corrected. Eventually, daughter cells containing only one type of plasmid arise

Fig. 1.3 Plasmid incompatibility. Mutual inhibition by replication control systems in a two-plasmid strain leads to plasmid segregation and the formation of single-plasmid cell lines.

(notably those in the IncF groups) contain more than one basic replicon, each with its own control system. What will happen when a plasmid containing two replicons (α and β) is challenged by an incoming plasmid containing replicon α? Although replicon α of the resident plasmid may be inactivated by inhibitor synthesized by the incoming plasmid, the plasmid will not be displaced if replicon β takes over the control of replication. The plasmids will be assigned to different incompatibility groups even though they may normally replicate from identical replicons and an unnecessarily large number of incompatibility groups will be invoked. An example of multiple replicons leading to apparently contradictory results was seen with pCG86 which contains two basic replicons: RepFIIA/FIC and RepFIB. In incompatibility tests pCG86 displaced IncFII plasmids but was compatible with IncFI. It was therefore assigned to group IncFII. Subsequently, however, the cryptic IncFI replicon was revealed when a spontaneous deletion derivative of pCG86 which had lost the RepIIA/FIC replicon proved incompatible with IncFI plasmids (Mazaitis *et al.*, 1981).

Minor genetic divergence between closely related plasmids may cause the inhibitor from one plasmid to interfere inefficiently with replication of another. This weakens the incompatibility reaction and can lead to difficulties in the interpretation of test results. Further confusion may arise because point mutations which simultaneously affect the inhibitor of replication and its target can change the plasmid's incompatibility behaviour radically. Studies on the control of ColE1 replication (see chapter 3) provide a remarkable example. The ColE1 replication inhibitor is a short, untranslated transcript (RNA I) whose target is the much longer RNA II which is processed to form the replication primer. The coding region for RNA I lies within that for RNA II (on the opposite strand) and RNA I exerts its inhibitory effect by base pairing to RNA II. Mutations which alter RNA I simultaneously make a complementary change in the target region of RNA II. Tomizawa and Itoh (1981) reported the isolation of a set of ColE1-like plasmids with altered incompatibility properties (*inc*). The mutations responsible all mapped to the RNA I–RNA II overlap region. Although they did not prevent binding between RNA I and RNA II from the mutant plasmid they reduced the affinity of the mutant RNA I for wild-type RNA II. A remarkable observation was that one of the mutant plasmids, with just two base changes in the overlap region, was completely compatible with the wild-type. In other words, just two base changes in a plasmid of several thousand base pairs were sufficient to create a new incompatability group.

In response to concern that multiple replicons and bifunctional mutations in inhibitor–target coding regions can lead to misleading conclusions from incompatibility tests, Couturier *et al.* (1988) have

developed a scheme of classification based on replicon typing. They
have established a bank of cloned replicons which are used as probes
to search for homology in the test plasmid. This affords a more direct
test of the relatedness of replicons because it detects the presence of
multiple replicons and will identify a replicon irrespective of whether
it functions in a particular plasmid environment.

1.4 The structure and organization of plasmids

1.4.1 Topological considerations

The overwhelming majority of *E. coli* plasmids are negatively
supercoiled circles of double-stranded DNA (Fig. 1.4). Supercoiling
influences a range of fundamental life processes including DNA
replication, recombination and transcription and is exploited in the
majority of plasmid purification techniques. The importance of this
aspect of plasmid structure warrants a brief detour into the realm of
DNA topology. A user-friendly treatment of this subject can be found
in Travers (1993) and a more advanced analysis in Boles *et al.* (1990).

The two strands of a linear DNA duplex can be completely
separated by heating to a temperature sufficient to melt the hydrogen
bonds between complementary bases. If the molecule is circular,
however, the strands cannot be fully separated because they are

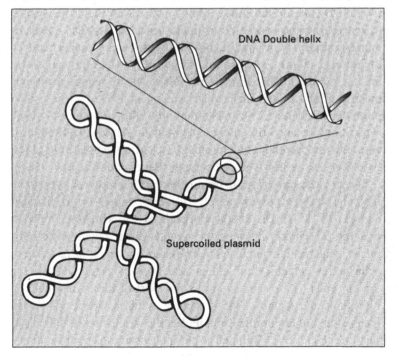

DNA Double helix

Supercoiled plasmid

Fig. 1.4 Plasmid
structure. The majority of
plasmids identified to date
are negatively supercoiled
circles. The plasmid
illustrated has four arms
or lobes of supercoiled
DNA. The size and
number of lobes varies
through writhing of the
DNA helices like worms in
a bucket.

linked not only by hydrogen bonding but also by the intertwining of the DNA backbones. The latter is known as the twist (Tw) component of linkage between the strands. In the absence of supercoiling (when the molecule is said to be relaxed), there is one link for each complete turn of the double helix (i.e. for every 10.5 bp of B-form DNA). *In vivo*, however, plasmids are not relaxed but supercoiled, and this alters the linking number. Supercoiling is introduced by the action of DNA gyrase, a type II topoisomerase (Reece and Maxwell, 1991) at the cost of ATP hydrolysis. A second enzyme, topoisomerase I, cannot introduce supercoiling but relaxes plasmid DNA by the ATP-independent removal of supercoils. The antagonistic action of gyrase and topoisomerase I maintain a constant level of supercoiling. An overview of topoisomerases and their role in the cell is provided by Luttinger (1995).

Two topologically distinct types of supercoiling are possible. In principle supercoiling can be either positive or negative, although the majority of circular DNA molecules in bacteria are negatively supercoiled. As a thought experiment we can imagine introducing negative supercoils into a plasmid by cutting both strands of the circular DNA molecule and rotating one of the cut ends in the *opposite* sense to the twist of the double helix (i.e. a left-handed rotation) before repairing the break. Although this is not the actual reaction mechanism of DNA gyrase, it achieves the same result. This process may to some extent unwind the double helix but it has an additional effect which can be seen when a length of rubber tubing is twisted. The axis of the rubber tubing or the DNA molecule writhes in space, crossing over itself at points called nodes (Fig. 1.5a). In topological terms, the introduction of negative supercoiling reduces the number of links between the strands of the DNA (known as the linking number). This favours processes which require transient melting of the double helix (e.g. binding of RNA polymerase to promoter sequences, recombination and the initiation of DNA replication) because the energy required to achieve it is reduced. Formally the overall linking number (Lk) is the algebraic sum of the twist (Tw) and writhe (Wr) components (Bauer and Vinograd, 1968).

Recent years have seen increasing interest in groups of hyperthermophilic archebacteria which live at temperatures up to, or even above, 100 °C. The DNA of these organisms resists melting, implying that the association between the strands must somehow be reinforced. The introduction of *positive* supercoiling would be one way to achieve this. Imagine cutting a circular DNA molecule and rotating the cut ends in the same sense as the twist of the double helix (a right-handed rotation). This process introduces writhe which increases the linking number and makes the strands more difficult to separate (Fig. 1.5b). Although a plausible scenario, positive super-

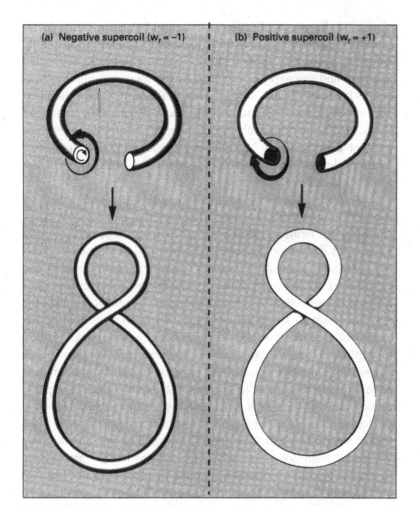

Fig. 1.5 Plasmid supercoiling. We can imagine introducing supercoiling into a circular plasmid by making a double strand break and twisting one of the cut ends before resealing the molecule. (a) If the twist (outer arrow) is of the opposite handedness to the double helix (inner arrow) the result is negative supercoiling. (b) If the imposed twist is of the same handedness as the double helix, positive supercoiling results.

coiling of plasmids in thermophilic organisms has yet to be demonstrated *in vivo*. Nevertheless, reverse gyrase (a topoisomerase which can introduce positive supercoils) has been isolated from hyperthermophiles (Bouthier de la Tour *et al.*, 1990), and viral genomes extracted these organisms are positively supercoiled.

1.5 The preparation of plasmid DNA

1.5.1 Caesium chloride density gradient centrifugation

It is not my intention to add to the many comprehensive descriptions of methods for the purification of plasmid DNA (see: Hardy, 1987; Grinsted and Bennett, 1988; Sambrook *et al.*, 1989, *inter alia*). It is, however, instructive to note how particular aspects of plasmid structure and topology are exploited by the most common purification techniques. The isolation of plasmid DNA presents an interesting problem. Plasmids constitute only one or two per cent of the total cellular DNA and chemically are indistinguishable from the chromosome of their host cell. Techniques for plasmid purification must therefore depend upon differences between the higher-order structures of plasmid and chromosome.

Shear forces associated with cell lysis and the early stages of purification invariably fragment the bacterial chromosome while plasmids, by virtue of their smaller compact structures, remain intact. This distinction proves extremely useful in plasmid purification. Originally plasmid preparation was a time-consuming process, culminating in the separation of plasmid and chromosome on a caesium chloride–ethidium bromide density. On such a gradient the DNA migrates to the point where its density equals that of the surrounding solution. In the absence of ethidium bromide this fails to separate plasmid from chromosome, except in unusual cases where they differ significantly in G + C content. Ethidium bromide is a planar molecule which, as it inserts between the DNA base pairs, tends to unwind the double helix. In linear DNA or nicked circles the linking number is not fixed because the ends are free to rotate. This makes unwinding less energetically demanding for chromosome fragments than for intact plasmids. Consequently chromosome fragments bind more ethidium bromide than plasmids and the two species equilibrate at different positions on the density gradient.

1.5.2 Rapid plasmid purification

At the end of the 1970s, molecular biologists were freed from density-gradient drudgery by the advent of rapid plasmid preparation methods. Like the density gradients they began to replace, the new methods exploited the topological differences between plasmid circles and linear chromosomal fragments. When the hydrogen bonds which link the complementary strands of circular plasmid DNA are broken either by heating (Holmes and Quigley, 1981) or by alkaline pH (Birnboim and Doly, 1979), the strands remain closely associated because they are linked by the intertwined backbones of the double helix. In contrast, the strands of linear or nicked DNA are free to to separate completely. If a mixture of denatured plasmid and chromosomal DNA is renatured rapidly (by cooling or restoration of neutral pH) the fidelity of reassociation differs substantially for the two species. The renaturation of plasmid circles is rapid and accurate because the strands are already in close physical proximity. Linear molecules generated by random shearing of chromosomal DNA renature less accurately, forming networks of DNA which can be removed from the lysate by centrifugation, together with denatured protein and RNA. Plasmid DNA remains in solution and can be precipitated with alcohol.

1.5.3 Gel electrophoresis of plasmid DNA

Agarose gel electrophoresis is often used to analyse plasmid DNA. Its rate of migration is determined by both size and physical structure. Electrophoresis of a plasmid preparation containing a variety of multimeric forms (Fig. 1.6a & b) demonstrates that the rate of migration of supercoiled molecules is inversely proportional to the log of molecular weight. This relationship can be used to estimate the size of unknown plasmids but comparisons must be made between

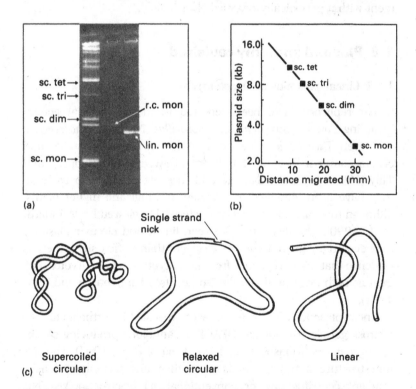

Fig. 1.6 Gel electrophoresis of plasmid DNA. (a) Agarose gel electrophoresis separates plasmid DNA by size and shape. Supercoiled (sc.) forms run fastest by virtue of their compact structure. Relaxed circular (r.c.) and linear (lin.) forms run more slowly because their open structures experience more resistance passing through the gel matrix. (b) For plasmids of the same physical state, the distance migrated is proportional to the log of plasmid size. (c) Structural differences between supercoiled, relaxed and linear plasmids account for their different rates of migration.

equivalent topological forms because supercoiled plasmids migrate much more rapidly than nicked or linear molecules (Fig. 1.6c). Fresh preparations of plasmid DNA are predominantly supercoiled but if there is any nuclease contamination in the sample, the introduction of single or double strand breaks can lead to the accumulation of relaxed and linear forms.

Until the mid-1980s agarose gels could not easily resolve molecules of more than 20 kb but the development of orthogonal field pulsed gel electrophoresis allowed resolution up to 2000 kb and led, amongst other things, to the discovery of giant linear plasmids in *Streptomyces* (Kinashi *et al.*, 1987). The technique was developed originally by Schwartz and Cantor (1984) who used it to separate intact yeast chromosomal DNA. A disadvantage of the original method was the requirement for complicated gel apparatus to deliver perpendicular electric fields. A later modification (Carle *et al.*, 1986) overcame this problem by using conventional electrophoresis equipment with a periodically reversed electric field.

1.6 Plasmid anatomy revisited

1.6.1 Linear plasmids of *Streptomyces*

Linear replicons have been reported in many bacterial genera including *Borrelia*, *Streptomyces*, *Thiobacillus*, *Nocardia*, *Rhodococcus*, and even *Escherichia*. The growing list of linear plasmids and chromosomes in bacteria has been reviewed by Hinnebusch and Tilly (1993). Indeed, linear plasmids are not confined to bacteria; they have been identified in a variety of fungi and higher plants, although most are still poorly characterized (reviewed by Meinhardt *et al.*, 1990). The distinction between linear and circular plasmids may eventually prove to be less clear-cut than at first sight; there is evidence that many replicons from prokaryotes and eukaryotes may exist as both linear and circular isomers (see Hinnebusch and Tilly, 1993).

Linear plasmids of over 300 kb were revealed by orthogonal field agarose gel electrophoresis (OFAGE) in antibiotic-producing strains of *Streptomyces* (Kinashi *et al.*, 1987; Kinashi *et al.*, 1993). The size and structure of these plasmids makes them undetectable by density gradient centrifugation or conventional gel electrophoresis. The best-characterized of these plasmids, pSR1, has terminal inverted repeats of 80 kb (Kinashi and Shimaji-Murayama, 1991). Smaller linear plasmids also contain terminal inverted repeats but these are generally less than 1 kb; the 17 kb plasmid pSLA2 from *S. rochei* has inverted repeats of 614 bp (Hirochika *et al.*, 1984).

1.6.2 Linear plasmids of *Borrelia*

Linear double-stranded plasmids with covalently closed ends have been studied extensively in the genus *Borrelia* (Barbour and Garon, 1987a). This group of host-associated spirochetes includes *B. burgdorferi*, the causative agent of Lyme disease and *B. hermsii* which is responsible for tick-borne relapsing fever in North America. Linear plasmids ranging from 15 to 200 kb are distributed widely among *Borrelia* species. Although the genetic composition of the majority of these linear replicons is unknown, the genes for major outer surface proteins of *B. burgdorferi* and *B. hermsii* have been found on linear plasmids (Barbour and Garon, 1987b, Plasterk *et al.*, 1985). The copy numbers of 49 kb and 16 kb plasmids have been estimated at only one per chromosome, implying a need for close control over their replication and partitioning (Hinnebusch and Barbour, 1992).

Intriguingly, the *Borrelia* chromosome migrates as a linear molecule of only 950 kb (Ferdows and Barbour, 1989), compared to the 3500 kb circular chromosome of *E. coli*. This, together with the low copy numbers of the linear replicons and reports that conventional circular double-stranded plasmids also occur in *Borrelia* (Hinnebusch and Barbour, 1992) has led to the suggestion that the linear plasmids might be regarded more properly as mini-chromosomes. The problem of determining where plasmids stop and chromosomes begin is not confined to *Borrelia*. Several genera of non-enteric Gram-negative bacteria contain huge plasmids. For example, *Rhizobium meliloti* harbours 'symbiotic megaplasmids' of 1.4 and 1.7 Mb in addition to a main chromosome of 3.4 Mb. Since these elements cannot be cured it may be more appropriate to classify them as chromosomes.

1.6.3 Linear plasmid telomeres

Problems associated with life as a linear replicon include how to achieve complete replication and how to protect ends from the ravages of exonucleases. In eukaryotes the problems are solved by the addition of telomere repeats by reverse transcription of a short RNA incorporated in the telomerase. Prokaryotic linear replicons characterized so far exhibit two different telomere structures exemplified by plasmids found in *Borrelia* and *Streptomyces* (Hinnebusch and Tilley, 1993).

A 16 kb linear plasmid from *B. burgdorferi* consists of a single polynucleotide chain which is fully base paired except for a short single-stranded hairpin loop at each end (Fig. 1.7a). The terminal sequence has homology with a 49 kb plasmid from the same species

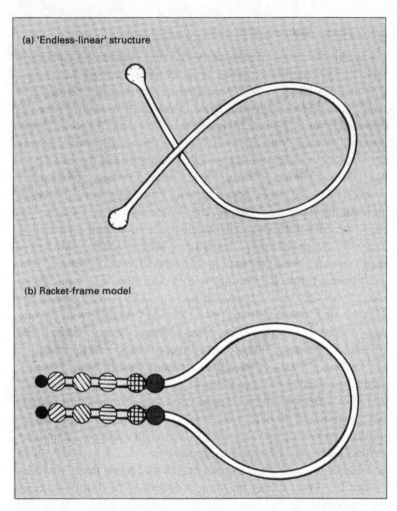

(a) 'Endless-linear' structure

(b) Racket-frame model

Fig. 1.7 End effects: telomeres of linear plasmids. (a) The 'endless linear' structure of plasmids found in *Borrelia* species consists of a double-stranded linear DNA molecule with single-stranded loops closing the ends. (b) The racket-frame structure proposed for linear double-stranded *Streptomyces* plasmids. The terminal regions of the molecule are proposed to associate through the action of juxtaposition proteins, forming the racket handle. The termini are protected by covalently attached proteins.

(Hinnebusch *et al.*, 1990) and similarities to the telomeres of other linear double-stranded replicons, including the iridovirus agent of African swine fever. The observation that this virus and at least one species of *Borrelia* share a common tick vector prompted the suggestion that the linear plasmids of *Borrelia* may have originated by horizontal transfer between kingdoms (Hinnebusch and Barbour, 1991).

Linear plasmids from the *Streptomyces* display a telomere organization which is also seen in adenoviruses, virtually all eukaryotic plasmids and some prokaryotic phages. The plasmids exhibit terminal inverted repeats with protein attached covalently to the 5' end of each DNA strand. Within the terminal repeats are multiple short regions of palindromic symmetry. It has been suggested that these bind proteins and bring together the plasmid termini in a 'racket frame' structure (Fig. 1.7b; Sakaguchi *et al.*, 1985).

2: The Unity of Plasmid Biology

2.1 Essential plasmid functions

A bewildering diversity of structures and phenotypes is illustrated in chapter 1 and the reader may fear that our exploration of plasmid biology will be no more than an exercise in molecular stamp collecting. We can avoid the perils of philately, however, if we concentrate on those aspects of biology which plasmids have in common. Plasmids are autonomous elements which have achieved independence by taking control of the means of production and of distribution. In biological terms these are the crucial processes of replication, inheritance and dissemination which are essential for the the survival of all plasmids. After a brief introduction in this section, they will be discussed in detail in succeeding chapters.

2.1.1 Persistence

For a plasmid to persist in a population of growing cells, its replication rate must match the division rate of its host. In practice, modulation of the frequency at which replication initiates maintains a constant plasmid concentration (described for convenience, rather than accuracy, as plasmid copy number). Mechanisms of replication control will be considered in detail in chapter 3.

The control of replication alone is insufficient to ensure that the plasmid persists. For low copy plasmids in particular, an additional function is required to ensure that at least one plasmid copy is received by each daughter cell. This is the role of active partition systems which distribute plasmids at cell division in a process analogous to the distribution of eukaryotic chromosomes during mitosis. Active partitioning is common only among low copy number plasmids; random distribution is probably sufficient for high copy plasmids because the frequency at which plasmid-free daughters arises is very low (less than 10^{-6} per cell division when the copy number is greater than 20 per dividing cell). The problem of plasmid inheritance will be considered in detail in chapter 4.

2.1.2 Proliferation

Although essential for plasmid maintenance, high-fidelity replication and distribution are insufficient to explain the ubiquity of plasmids in natural populations. Even very occasional failures of replication or distribution will result in a slow but inexorable decline in the number of plasmid-containing cells. A notable characteristic of plasmids which helps combat this decline is their capacity for cell-to-cell transfer. In population genetics terminology, plasmid fitness has both a vertical component (persistence from one generation to the next), and a horizontal component (proliferation) which reflects the ability to colonize new hosts. In chapter 5 we look in detail at this important aspect of plasmid ecology and review cell-to-cell transfer of plasmids by a variety of mechanisms including conjugation, transduction and transformation.

It seems likely that in nature plasmid proliferation is most often by conjugation (Amabile-Cuevas and Chicurel, 1992). Because conjugation involves simultaneous replication and transfer, it provides a way for the plasmid to out-replicate its host and increase the total number of copies in the population. If cells which receive the plasmid are favoured by natural selection, it will spread rapidly. A striking illustration of this is the appearance world-wide of antibiotic-resistant bacteria which will be described in detail in chapter 6. In the 1950s it became apparent that combinations of resistance arising in one species could be transferred as a unit to other species in the same environment. The increased fitness associated with the acquisition of resistance plasmids led to their rapid spread in hospital environments. The way in which R plasmids crossed species barriers was an eloquent demonstration that a theory of evolution by mutation and selection was no longer adequate. Where plasmids are involved, evolution proceeds by quantum leaps as species acquire ready-made sets of genes adapting them at a stroke to new environments.

2.1.3 Cryptic plasmids

Plasmids have often been classed as symbionts because, in return for their maintenance, they appear to confer a selective advantage on their host. Antibiotic resistance plasmids are often cited as evidence, but this view is dangerously naïve because the advantage is limited to certain, rather atypical environments. Furthermore there are many plasmids which offer nothing in return for the metabolic load which they impose. How could such cryptic parasites become established in bacterial populations? A trivial possibility is that many of these plasmids increase host fitness in some way which has not yet been recognized. If genuinely cryptic, plasmid loss through segregational instability must be at least balanced by horizontal spread through conjugation, transduction or transformation. Mathematical models designed to test this hypothesis take account of the rate at which plasmid-free cells arise, the effect of non-selected plasmids on cell growth and the frequency of horizontal transfer. The results of these analyses have been far from unanimous. An early report suggested that non-selected plasmids might indeed proliferate by conjugation alone (Stewart and Levin, 1977), but more recent studies have concluded that plasmids must confer a selective advantage if they are to survive (Simonsen, 1991; Gordon, 1992).

If, as recent analyses would have us believe, cryptic plasmids cannot proliferate by conjugation alone, how do they become established? Theoretical proofs of the impossibility of insect flight warn that any model building exercise, and the conclusions drawn from it, must be treated with caution. Estimates of key parameters are often made in the laboratory and may not accurately reflect the behaviour of cells and plasmids in their natural environments. Conjugation frequencies estimated in liquid culture are likely to be significantly lower than those on solid surfaces (Bradley *et al.*, 1980) and in biofilms (Costerton *et al.*, 1987). A further complication, often ignored in quantitative models, is that the frequency of conjugation is variable. Conjugal transfer is tightly regulated and, for most plasmid–host combinations, is a rare event. However, immediately after a cell has received a plasmid there is transitory derepression of the transfer system and a high probability that the plasmid will be passed to a plasmid-free neighbour. The result is an epidemic spread of the plasmid; a phenomenon which Lundquist and Levin (1986) have proposed may be sufficient to maintain the plasmid in the absence of direct selection for increased host fitness.

The effect (cost or benefit) of a plasmid on host fitness is invariably assumed constant in mathematical analyses, but may in reality vary with growth rate, nutrient availability and the length of the plasmid–host association. Bouma and Lenski (1988) found that *Escherichia coli* B transformed with pACYC184 was initially less fit than plasmid-free cells. After 500 generations, however, curing cells of the plasmid actually reduced their fitness. Coevolution of host and plasmid was shown to have resulted from changes in the host genome rather than the plasmid. As a result of this analysis, Bouma and Lenski caution against uncritical acceptance of the 'excess baggage' hypothesis that the deliberate release of organisms containing recombinant DNA into the environment is safe because plasmid load will prevent their unintended spread.

2.2 Plasmids and prokaryote evolution

The concept of the species as a reproductively isolated unit is fundamental to our understanding of eukaryote evolution. Hand-in-hand with this goes the belief that genetic information is transmitted vertically (from parent to offspring) but never horizontally. Among prokaryotes, horizontal gene flow is ubiquitous and is most commonly and efficiently mediated by plasmids. A consideration of the mechanisms of horizontal plasmid transfer is postponed until chapter 5. Here we will consider how a combination of horizontal transfer and fluidity of genome organization means that plasmids constitute a pool of universally available prokaryote genes which show scant regard for the labours of microbial taxonomists.

2.2.1 Changing and yet staying the same

Bacterial species vary little over enormous periods of time. *E. coli* and *Salmonella typhimurium* exhibit closely related physiology and genetic organization, yet molecular studies of rRNA show that they last shared a common ancestor 120–160 million years ago; the time of the appearance and diversification of mammals (Ochman and Wilson, 1987). Against this background of extraordinary genetic stability, it is perhaps surprising that bacteria can respond to environmental change with impressive rapidity. The rapid rise of antibiotic resistant strains is just one example of this. The solution to the paradox lies in the division of labour between chromosomal and plasmid genes and the different patterns of inheritance displayed by the two classes of replicon. The chromosome ensures genetic stability while plasmids provide opportunities for experimentation and rapid change.

Plasmid genes move horizontally by transduction, transformation and conjugation both between and within species (Fig. 2.1). Broad host range plasmids move freely between Gram-positive and Gram-negative organisms and are even capable of transfer between kingdoms from bacteria to yeast (Heinemann and Sprague, 1989; Mazodier and Davies, 1991). Of course the transfer of plasmid-borne genes between species is not entirely free. It is relatively rare and is opposed by host-encoded restriction systems and the diversity of signal sequences for replication and gene expression. Nevertheless, some plasmids encode anti-restriction proteins (Delver *et al.*, 1991) and rates of plasmid transfer which seem low when measured in the laboratory may have dramatic consequences on an evolutionary timescale.

In view of the many opportunities which exist for cell-to-cell transfer, it is probably inappropriate to think of plasmids as specific to individual bacterial species. A more realistic view may be that plasmid genes are delocalized over all bacterial species in a manner reminiscent of the π-electrons in the benzene ring which are not permanently associated with any one covalent bond. A single plasmid capable of horizontal transfer may enable a range of host species to flourish in a particular environment. This was illustrated by an

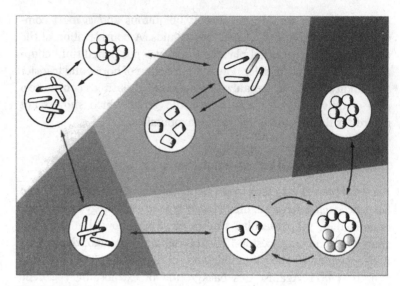

Fig. 2.1 Gene flow among bacterial species. While eukaryotes exist in genetic isolation, plasmid-mediated gene flow blurs the definition of bacterial species. Opportunities exist for gene transfer between species in a common environment and sometimes between organisms which normally inhabit different environments (for example, enteric organisms ejected in faeces may exchange genes with the endemic soil flora and subsequently re-invade the gut of grazing animals).

outbreak of multiple drug resistant infections in a Melbourne hospital caused by strains of *Klebsiella, E. coli, Serratia marcescens, Proteus mirabilis and S. derby*. Identical conjugative R plasmids were found in at least one representative of each genus (Davey and Reanney, 1980).

2.2.2 Plasmids as libraries of genetic information

If we count only chromosomal genes, the pool of information available to a bacterium seems small. If, instead, we consider all plasmid genes to be available to all bacteria, the pool of information available to each individual species exceeds that available to a simple but genetically isolated eukaryote such as yeast (Reanney, 1976). This has important consequences for bacterial evolution. In an ecosystem where sufficient genetic interconnections exist, the size of the entity on which selection acts will far exceed the size of the gene pool of the same species under laboratory conditions. If the eukaryote genome resembles the well-stocked book shelves of a private individual, then prokaryotes have relatively few volumes of their own but they have access to an impressive lending library of plasmid borne-genes. It may take longer to obtain a book from the library than from one's own shelves but the library offers an enormous supply of titles at very low cost to the individual.

2.2.3 The division of labour between plasmid and chromosome

The bewildering variety of plasmid-encoded phenotypes raises the question whether genes on plasmids have any common features. Even if we exclude those functions concerned exclusively with plasmid replication and maintenance, plasmids certainly do not represent a random sample of the bacterial gene pool. Eberhard (1990) has proposed that plasmid genes promote the survival of bacteria in rare and exotic environments while those involved in mundane housekeeping functions remain on the chromosome. He has advanced the local adaptation hypothesis to explain this asymmetric distribution. Eberhard argues that plasmid-borne genes confer adaptation to environments which occur sporadically over time or space. Whenever appropriate conditions arise, genes which increase host fitness spread most rapidly when in genomes capable of horizontal transmission. To illustrate this argument, consider a population of bacteria in an antibiotic-containing environment where genes encoding resistance occur at low frequency on both plasmids and chromosomes. The plasmid gene has an advantage over a chromosomal equivalent if bacteria lacking the resistance gene attempt to colonize the environment. These newcomers (assuming they are receptive to plasmid transfer) benefit from the ready availability of the plasmid-borne genes. If the propagation of the plasmid by horizontal spread is greater than its rate of loss due to segregational instability, the plasmid resistance gene will propagate more rapidly than its chromosomal counterpart. As a direct consequence of its mobility, the plasmid gene also experiences a greater variety of genetic backgrounds than the chromosomal version, some of which may enhance its effect and increase host fitness. This constant reshuffling scenario will be familiar to evolutionary biologists as one of the arguments for the ubiquity of sexual reproduction.

2.2.4 Do plasmids ever carry essential genes?

Most plasmid-encoded traits are optional in the sense that they only increase host fitness in atypical and often transient environments. Sometimes, however, a plasmid appears to be an essential part of the host genome. Examples include plasmids encoding virulence factors which are present in all isolates of an obligate enteric pathogen (Brubaker, 1985). The role of plasmids in insect pathogens has attracted considerable interest because of the potential exploitation of these organisms in pest control. *Bacillus thuringiensis* subspecies contain a substantial proportion of their genetic complement as plasmid DNA. The plasmids appear to be involved in both the production and regulation of protoxins, some encode bacteriocins and others are cryptic. A broad size distribution is seen among these plasmids; subspecies *berliner* and *kurstaki* have a cluster of plasmids around 4–6 MDa, several larger than 30 MDa and some over 100 MDa (Aronson *et al.*, 1986). Attempts to cure *B. thuringiensis* plasmids have met with only limited success. Strains thought to be plasmid-free were found to contain a 130 MDa plasmid of unknown origin and these 'cured' strains grew less well than their plasmid-containing ancestors. They show defective spore coats and growth yields are reduced by 10–20%, suggesting that one of the plasmids might provide supplementary growth factors when nutrients become limiting.

2.3 Structural fluidity of plasmid genomes

A high degree of structural fluidity assists plasmids in their role as distributors of genes among prokaryotes. Recombination and transposition reshuffle plasmid genomes, endlessly generating novel genotypes. Regions of plasmid or chromosomal DNA become mobile when flanked by insertion sequences, and transposition can lead to large-scale plasmid rearrangements including fusion with other plasmids or integration into the chromosome. The resulting structural fluidity means that it is probably unwise to view any individual plasmid as more than a transient association of information modules drawn from a universally available gene pool.

An example of structural fluidity is provided by pHH1457 (a large conjugative plasmid from *K. aerogenes*) which spontaneously deletes 25% of its DNA at a frequency of 0.25% per generation. The deletion is associated with a loss of conjugative ability and the majority of resistance genes, although clinically important resistance to gentamycin is retained (Blenkharn and Hughes, 1982). Such instability is not uncommon; changes in patterns of antibiotic resistance in pathogenic organisms correlate frequently with the acquisition and loss of R plasmid genes (Chau *et al.*, 1982). Conjugative plasmids of the RepFIIA family appear to have exchanged modules encoding incompatibility determinants (Lopez *et al.*, 1991) while plasmids of Gram-positive bacteria have exchanged casettes encoding mobilization genes, replication proteins and resistance determinants (Projan and Moghazeh, 1991).

2.3.1 Transposons and plasmid structure

Transposons play an important role in the modification of plasmid structure. Genes can be imported from the chromosome on a mobile element and subsequent movement of the transposon may result in gross structural rearrangements of the plasmid genome.

The most obvious change associated with transposition is the insertion of the element into a new target sequence; a process which accounts for 5–10% of spontaneous mutations. If the target and donor sites are within the same molecule, elements which replicate during transposition cause deletion or inversion of adjacent sequences. Under some circumstances similar rearrangements are also associated with movement of non-replicative composite transposons. When transposition is replicative, movement to a target in another molecule may lead to cointegration (fusion) of donor and target replicons. Mechanisms of transposition and their associated rearrangements have been reviewed in depth by Grindley and Reed (1985).

2.3.2 Recombination and plasmid structure

Although transposon-mediated events have been of undoubted significance in the evolution of antibiotic resistance plasmids, structural variation and rearrangement may also arise through recombination. RecA-dependent homologous recombination requires a DNA match between recombining sequences of at least 25–30 bp. Chance repetition of 25 bp sequences is very infrequent (roughly once every 10^{15} bp) but transposons and insertion sequences act as mobile regions of homology between which recombination often occurs. Plasmid fusion or integration of plasmid DNA into the chromosome is a result of intermolecular recombination. Inversions and duplications are generated by intramolecular recombination between sequences in direct and inverted repeat, respectively.

RecA-independent (sometimes called illegitimate) recombination between 5 and 10 bp repeats is responsible for a variety of spontaneous rearrangements including deletions, inversions, duplications and fusions. They are important because repeats of this length occur by chance every few kilobases. DNA gyrase-mediated events belong in this category. DNA gyrase is a type II topoisomerase which introduces negative supercoils into covalently closed DNA with associated hydrolysis of ATP. Gyrase binds to a consensus sequence where it introduces a double-strand break in the DNA. In a transient intermediate, the cut ends of the DNA are linked covalently to an amino acid residue on the topoisomerase. Normally, an intact length of DNA is passed through the break which is then resealed. At low frequency, subunit exchange between pairs of gyrase-binding site complexes can result in deletion or inversion of the intervening DNA. Illegitimate recombination is also associated with the initiation and termination of replication and copy-choice errors in replication where slippage of the replication apparatus generates deletions (Berg, 1990).

2.4 Where do plasmids come from?

2.4.1 Plasmids and bacteriophages

It is not clear whether plasmids are the relics of bacteriophage which lost the ability to package their genome and escape from the host or whether phage evolved from plasmids by the acquisition of these functions. These possibilities are not, of course, mutually exclusive and it is indisputable that the life-styles of plasmids and phage are closely related. Both phage and plasmids are parasites in the sense that they have no means of reproduction which is independent of their host and do not unconditionally increase the likelihood of their host's survival and reproduction (Levin and Lenski, 1983). Often the distinction between phage and plasmid is blurred. Thus the temperate phage P1 carries genes involved in copy number control and active partition and can exist indefinitely as a unit copy prophage, indistinguishable from any respectable low copy number plasmid. Spontaneous deletions of λ which retain little beside the bacteriophage replication origin persist as plasmids. The genome of M13, both loved and loathed by molecular biologists as the original vector for Sanger DNA sequencing, has two distinct states. The infectious particle contains a single strand of DNA which is converted to a double-stranded plasmid after infection and it is this plasmid form which serves as the template for synthesis of single-stranded phage genomes.

Imagine a nascent plasmid (or phage), newly emerged from the host chromosome and fortuitously containing an origin of replication. Ejected from the chromosomal nest, this fledgeling replicon must acquire some means of over-replication (i.e. replicating faster than the host) if it is to survive. Levin and Lenski (1983) envisage two ways in which this might be achieved. The path of 'niceness' involves the acquisition of genes which enhance host fitness (i.e. the evolution of a symbiotic relationship) while the alternative of infectious transmission requires the element to out-replicate its host by conjugal transfer or, as a phage, by lytic growth. Symbiosis is likely to be more important in low-density populations with few opportunities for horizontal transmission while in high-density populations, elements with effective mechanisms for horizontal transfer would be favoured. At intermediate densities or in populations with widely fluctuating density, an evolutionary compromise might permit both temperate phages and conjugative plasmids to flourish.

2.4.2 Parasites or symbionts?

The relationship between plasmid and host is enigmatic. Are

plasmids symbionts exchanging increased fitness for metabolic favours from the host or selfish DNA parasites which proliferate at the expense of the host without providing long-term benefit? Cryptic plasmids seem entirely parasitic but what of plasmids which confer resistance to heavy metals or antibiotics; are they not symbiotic? This does not necessarily represent a true symbiosis because the increase in host fitness is restricted to a restricted and often atypical environment. In nature, it is probably rare for bacteria to encounter significant concentrations of antibiotics or heavy metals. Resistance genes may serve as a 'loss leader' to establish the plasmid in a bacterial population. When conditions change and the plasmid-encoded phenotype is no longer advantageous, the nature of the plasmid–host relationship changes and the plasmid is transformed into a tenacious parasite.

Large plasmids encode a variety of systems which ensure their persistence irrespective of their effect on host fitness. Active partition ensures that plasmids are distributed to both daughter cells and plasmid dimers which would interfere with the process are converted to monomers by site-specific recombination. Even if a plasmid-free cell should arise due to failure of replication or partition, the unfortunate cell is likely to be killed by the translation or activation of toxic polypeptides encoded by its former resident (post-segregational killing functions are discussed in Chapter 4). The plasmid which may initially have become established in the population by increasing host fitness is able to persist when conditions change and the plasmid metabolic load, no longer balanced by any benefit, begins to reduce the fitness of its host. The spontaneous deletion of antibiotic resistance genes from R factors may be a strategy by which the established parasitic plasmid minimizes metabolic load and maximizes host fitness.

2.4.3 Plasmids, bacteriophages and transposons

Naomi Datta (1985) proposed the idea that plasmids, transposons and bacteriophages constitute a super-family of organisms whose members are parasitic and capable of both vertical and horizontal transmission (Table 2.1). Horizontal transmission is by replicative transposition, plasmid conjugation or lytic growth of bacteriophages. The super-family might even be extended to include retrons; those enigmatic elements which are the source of multicopy single-stranded DNA in diverse bacterial genera (reviewed by Inouye and Inouye, 1992). Although some of the elements in the super-family appear to have a symbiotic relationship with their host (e.g. plasmids and transposons which encode antibiotic resistance or lysogenic phage which protect their host from lytic infection), Datta steadfastly

Table 2.1 A superfamily of molecular parasites.

Element	Mode of transmission	
	Vertical	Horizontal
Plasmid	Active partition or random distribution	Conjugation, transduction or transformation
Transposon	Passive transmission as part of an autonomous replicon	Replicative transposition
Temperate bacteriophage	Transmitted as an integral part of a replicon (λ lysogen) or as an autonomous replicating element (P1 prophage)	Lytic growth
Virulent bacteriophage	?	Lytic growth

classes them all as parasites because symbiosis is not a necessary part of their existence. She admits that the unity of the super-family is defined as 'an act of faith' but the concept certainly makes it easier to reconcile the behaviour of elements which refuse to fit neatly into a more narrow classification. In addition to bacteriophages which behave as plasmids for part of their life cycles, a good example of an element with an identity crisis is bacteriophage Mu which proliferates by replicative transposition but is packaged and lyses the host like a bacteriophage.

3: Plasmid Replication and its Control

3.1 Essential components of replication control systems

3.1.1 The need for replication control

Autonomous replication is a fundamental plasmid characteristic. However the plasmid is not entirely a free agent; it must synchronize its replication with the growth and division of the host cell. To achieve stable coexistence with its host, each plasmid must replicate, on average, once every generation. A plasmid which replicates less than once a generation will suffer a decline in copy number until plasmid-free cells appear in the culture. Conversely, if the plasmid replicates more than once each generation, its copy number will rise and the metabolic load imposed upon its host will have an increasingly detrimental effect on viability. In this chapter we consider the principles underlying replication control and review what is known about the various mechanisms by which control is achieved in practice. A variety of examples have been chosen to illustrate the diversity of control systems but, in the interests of clarity, no attempt has been made to provide a complete catalogue of every system on which work has been published.

3.1.2 Basic principles of copy number control

Plasmids are maintained at a constant mean copy number (more strictly a constant concentration) under defined growth conditions. Copy number is buffered against disturbance and this is most clearly seen when a single plasmid is introduced by transformation or conjugation; normal copy number is rapidly established and subsequent deviations are corrected (Highlander and Novick, 1987). Such observations provided the basic evidence for the existence of replication control systems which, by sensing and correcting deviations, both define and maintain plasmid copy number (Pritchard 1978).

The basic requirements for effective copy number control are simple. The control circuit must ensure that the average replication rate is less than once *per* generation in cells with too many plasmids and more than once in cells with too few. The variety of possible relationships between copy number and replication frequency have been discussed by Nordström *et al.* (1984) and are summarized in

Fig. 3.1. Curve **A** represents a system where each plasmid replicates once every generation, irrespective of copy number. This system is unregulated because deviations in copy number (due for example to the uneven distribution of plasmids at cell division) remain uncorrected, resulting in an ever broader copy number distribution in the population. Chiang and Bremer (1991) have demonstrated this experimentally using derivatives of pBR322 with defective copy number control.

Curve **B** in Fig. 3.1 represents a control system in which the replication frequency is inversely proportional to copy number. Thus if the copy number falls to half the mean value, the replication frequency per plasmid doubles. Conversely, if the copy number doubles, the replication frequency is halved. Under these conditions, the replication rate per plasmid is variable but the total number of replication events per cell per generation is constant; a condition which has been demonstrated experimentally for plasmid R1 (Nordström *et al.*, 1984). Curve **C** represents another controlled system, but this time replication is switched on or off by very small changes in copy number. The shape of the step function means that above a critical copy number, replication is totally inhibited. This type of 'all or nothing' control is characteristic of plasmid F (Tsutsui and Matsubara, 1981) where replication is inhibited totally at twice the normal copy number. It is, of course, possible to imagine any

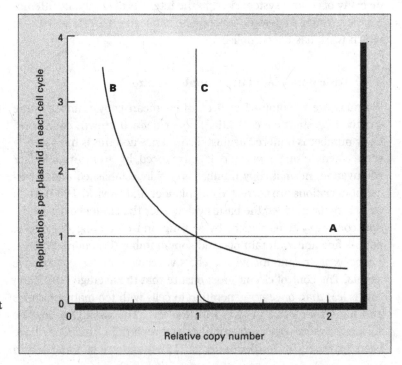

Fig. 3.1 The relationship between replication rate and plasmid copy number. Both hyperbolic (B) and step (C) relationships result in effective copy number control. When the replication rate is constant (A) copy number is uncontrolled. (From Nordström *et al.*, 1984.)

number of curves lying between A and C, each of which fulfils the replication rate criteria and therefore represents an effective control system.

3.1.3 The basic replicon

Although plasmid replication could potentially be regulated at the level of initiation, elongation or termination, there is overwhelming evidence for control over initiation and specifically over the initiation of leading strand synthesis. A starting point in most studies of replication control is to separate DNA sequences involved in initiation from other plasmid genes. In the majority of plasmids, the replication functions are clustered within a region of 1-3 kb known as the basic replicon. This was defined originally as the smallest portion of the plasmid able to replicate (Kollek *et al.*, 1978). Nordström (1985) has suggested a narrower and more useful description of the basic replicon as the smallest piece of DNA that replicates *with wild-type copy number*.

3.1.4 Replication origins

Replication initiates at a site known as the origin (*ori*). Origins were identified originally using electron microscopy to identify the source of replication 'bubbles'; an aesthetically pleasing although rather imprecise technique. An exact definition of the origin is the position at which the first deoxyribose base is added to the leading-strand RNA primer, but there seems to be some variability in the precise point of switching from RNA to DNA. It can occur at any one of three consecutive bases in plasmid ColE1 (Tomizawa *et al.*, 1977) and within either of two clusters of three or four closely linked bases at the origin of chromosome replication (Hirose *et al.*, 1983).

3.1.5 Multireplicon plasmids

Small, multicopy plasmids such as ColE1 typically contain a single basic replicon. Many large plasmids (such as those of the IncF incompatibility groups) contain multiple replicons, although often only one is active *in vivo*. Origin selection in multireplicon plasmids is determined by complex and poorly understood mechanisms and attempts to identify the primary origin are fraught with difficulty. The conventional approach involves linking restriction fragments to a drug resistance gene and screening for autonomous replication. This can be misleading, however, because secondary replicons may be activated when separated from the rest of the plasmid.

There are several independent pieces of evidence for interactions between replicons, including the early observation of Figurski *et al.*

(1979) that in an RK2-ColE1 fusion, the ColE1 origin was functional only when the copy number of the fusion was below five per cell (the copy number of RK2). More recently, Maas *et al.* (1989) found that in plasmids containing the RepFIIA/FIC and RepFIB basic replicons, challenging one replicon with an incompatible plasmid interfered with replication from the other. Working with multicopy plasmids in *Staphylococcus aureus*, Projan and Novick (1992) reported that in fusions between plasmids pT181 and pE194, the pT181 replicon was inhibited by the second-strand replication origin of pE194.

It is unclear why some plasmids contain multiple replicons while others have only one. It has been suggested that secondary origins are used by plasmids in exotic hosts. A change in origin usage is seen when NR1 is transferred from *Escherichia coli* to *Proteus mirabilis* (Warren *et al.*, 1978) but, on the other hand, all three R6K origins are used with equal frequency in *E. coli* (Kolter, 1981). Another possibility is that multiple replicons increase the effective host range of a plasmid by allowing colonization of bacteria in which the primary replicon is repressed by incompatibility with a resident plasmid. They may even provide an insurance policy against the potentially disastrous effect of transposon insertion within the replicon. A less ingeneous but perhaps more plausible suggestion is that replicon number and organization is not the result of natural selection (an assumption implicit in all of the above ideas) but is a consequence of continuous reshuffling of the plasmid gene pool. There is certainly considerable scope for homologous recombination between related transfer operons of the IncF plasmids. If the reshuffling hypothesis is true, plasmid organization would seem to have more in common with the apparent chaos of the eukaryote genome than the parsimonious organization of the bacterial chromosome.

The complexity of multireplicon plasmids makes it impossible to classify them into unique incompatibility groups. This is illustrated by difficulties experienced with the IncF incompatibility groups (Bergquist *et al.*, 1986). Three basic replicons have been identified among plasmids of the IncFI group: RepFIA, RepFIB and RepFIC. When each of these was used as a probe to determine its distribution among 40 plasmids of the IncF group, homology to IncFIC was found in all but one, homology to IncFIB in 25 and homology to IncFIA in 17. Data derived from the use of replicons as probes must be interpreted with care, however, as sequence homology may indicate a common evolutionary origin even when replicons have diverged sufficiently that they no longer interfere with one another's replication or are no longer active. An example is RepFIC which has a similar DNA sequence to the primary RepIIA replicon in IncFII plasmids such as R1, although plasmids containing these origins are compatible (Saadi *et al.*, 1987).

3.2 Strategies of replication control

3.2.1 Passive control

In principle, replication control could be achieved either by an active process, where a feedback loop links plasmid copy number to replication rate, or passively, where replication is limited by some external (i.e. host-imposed) constraint. In the latter case, replication might be limited by the concentration of an essential, host-encoded protein. Under constant growth conditions, the protein would reach a steady-state concentration which would determine the plasmid replication rate and hence its copy number. Although very simple, a passive system will correct copy number deviations since a fall in plasmid copy number at constant protein concentration will result in a higher replication rate *per* plasmid, while an increase in copy number will result in a decreased rate.

Despite their seductive simplicity, it is unlikely that passive mechanisms are of widespread importance in replication control. The rate-limiting factors would have to be incompatibility group specific because, by definition, compatible plasmids have independent control systems. There are more than 20 incompatibility groups in *E. coli* and the biochemical similarity of the host functions required for replication of all plasmids argues against the existence of a large number of discrete, rate-limiting factors. In the end, the strongest evidence against a major role for passive control is that all plasmids studied to date employ an active, plasmid-encoded system.

Before dismissing passive control completely, it is worth reflecting that although active control circuits have been shown to determine copy number in the laboratory, we know little about the situation in more stressful natural environments. Here, physiological constraints may reduce the maximum replication rate so severely so that copy number never rises to the level where plasmid control circuits limit replication (Thomas 1988). Passive control may also come into play in some highly unnatural situations. In copy number mutants where the primary control circuit is inactive, plasmid replication may eventually be limited by the availability of a host-encoded factor. Under these conditions, normal incompatibility relationships would break down and all plasmids should contribute to a plasmid pool of fixed size. Experimental evidence on this matter is equivocal. Chiang and Bremer (1991) studying the maintenance of pBR322 derivatives with inactive control circuits found that limiting host factors failed to provide a back-up to stabilize copy number. In contrast, Gelfand *et al.* (1978) reported that a mutant of ColE1 with elevated copy number and a deletion derivative of the same plasmid both comprised about 27% of total cellular DNA. Thus the copy number of the deletion derivative was higher than the original plasmid but the total amount of plasmid DNA was unchanged.

A direct approach to the question of plasmid carrying capacity was adopted by Ruby and Novick (1975) who established many compatible plasmids in a cell and asked whether the total copy number was always the sum of the copy numbers of individual plasmids. They concluded that the maximum plasmid space in *Staphylococcus aureus* is about 900 kb and rationalized their observation by speculating that organisms finding it useful to carry six or more compatible multicopy plasmids would presumably evolve some mechanism to avoid paying the 'metabolic tax' of 40% that additivity would demand.

3.2.2 Active control

In active systems, replication is controlled by the plasmid itself. There have been attempts in the past to propose active control systems based upon both positive and negative control (i.e. where the rate-limiting factor is a plasmid-encoded activator or repressor, respectively). However, Nordström (1985) has argued that any serious attempt to describe a positive system must invariably end in failure. To achieve a decreasing replication rate for increasing plasmid concentration, the only alternative to a negative control loop is for the replication rate to be set by a rate-limiting, host-encoded factor (i.e. passive control).

3.3 Model systems of replication control

3.3.1 The formulation of negative control models

Autorepressor and Inhibitor Dilution were two influential models of replication control which were proposed before there had been an extensive molecular analysis of any real system. They both describe active systems in which the replication rate is set by a plasmid-encoded repressor. In the Inhibitor Dilution model the repressor is produced constitutively and its concentration is proportional to plasmid copy number. In the Autorepressor model, repressor synthesis is subject to negative feedback and its concentration is constant. Both meet the necessary criterion that the replication rate *per* plasmid increases or decreases appropriately when the copy number deviates from the mean.

3.3.2 The Inhibitor Dilution model

The Inhibitor Dilution model describes a system in which replication is controlled by a plasmid-encoded inhibitor (Cop) whose concentration is proportional to plasmid copy number (Fig. 3.2). Close proportionality can most easily be achieved by constitutive synthesis of an unstable repressor.

In their original statement of the model, Pritchard *et al.* (1969) assumed that small changes in inhibitor concentration would have a large effect on replication rate. This resembles the step function represented by curve C in Fig. 3.1. As a cell grows, the volume increases and the the inhibitor concentration falls. This results in the temporary derepression of plasmid replication and an increase in copy number. More plasmid copies means more inhibitor genes, and the inhibitor concentration rises until it reaches the critical value

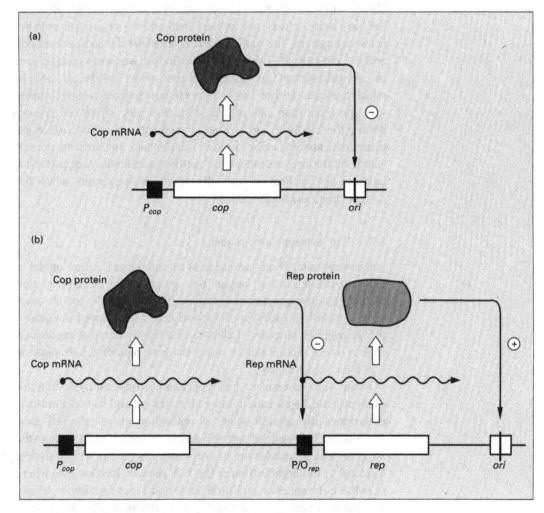

Fig. 3.2 Alternative realizations of the inhibitor dilution model. (a) An inhibitor protein (Cop) is synthesized by the plasmid and acts directly at the origin to inhibit replication. (b) The inhibitor (Cop) acts indirectly by preventing synthesis of an essential initiator (Rep) protein.

when replication is switched off. Further cell growth again dilutes the repressor and the cycle is repeated.

The replication control system of plasmid R1 operates in a manner similar to that described by the Inhibitor Dilution model. One difference is that the replication rate of R1 is inversely proportional to plasmid copy number (curve B in Fig. 3.1); an hyperbolic relationship rather than a step function. This was concluded from an experiment in which the copy number of an R1 derivative was elevated four- to six-fold and the plasmid replication rate monitored during its subsequent return to normal. Throughout this time, plasmid replication continued and in each cell the total number of replication events *per* generation was constant. Furthermore, the rate was equal to that observed in a steady-state culture (Nielsen and Molin, 1984). Because the copy number was initially very high and the replication rate *per* cell was constant, the replication rate *per plasmid* was less than one and the copy number declined. This finding was incorporated in a revised statement of the Inhibitor Dilution model (Pritchard, 1985) in which replication is not simply on or off, but changes in copy number vary the replication rate of individual plasmids. When cell growth dilutes the inhibitor, each plasmid replicates more than once *per* generation and the copy number climbs back towards its normal value. As the copy number (and hence the inhibitor concentration) increase, the replication rate falls again, reaching unity when the correct copy number is reached.

3.3.3 The Autorepressor model

A few years after the appearance of the Inhibitor Dilution model, a different active control system was proposed by Sompayrac and Maaloe (1973) who wanted to explain the control of *E. coli* chromosome replication. Although the model eventually proved unsuitable for its original purpose, it illustrates a mechanism of replication control which is employed by many elements including λ-*dv*, plasmid F and the P1 prophage.

In the Autorepressor model, replication is triggered by an initiator (Rep) protein. Rep is rate-limiting for initiation and its concentration determines the total number of replication events *per* cell each generation. The *rep* gene is part of an operon which also contains the *atr* gene which encodes an autorepressor (Fig. 3.3a). This is almost identical to the organization of the λ-*dv* replicon (section 6.6) where cI is the autorepressor and proteins O and P act together to trigger replication. In an alternative statement of the model, the functions of autorepressor and initiator are combined in a single bifunctional protein (Fig. 3.3b); an organization which is seen in the P1 prophage (section 6.7). The negative feedback loop provided by autorepression ensures that the concentration of Rep protein (and therefore the replication rate) is constant and independent of cell volume, growth rate or plasmid copy number. A constant replication rate per cell means that the rate per plasmid is inversely proportional to copy number; the relationship represented by curve B in Fig. 3.1.

3.3.4 The kinetics of control: oscillation and over-shoot

In the Inhibitor Dilution model, the inhibitor concentration is

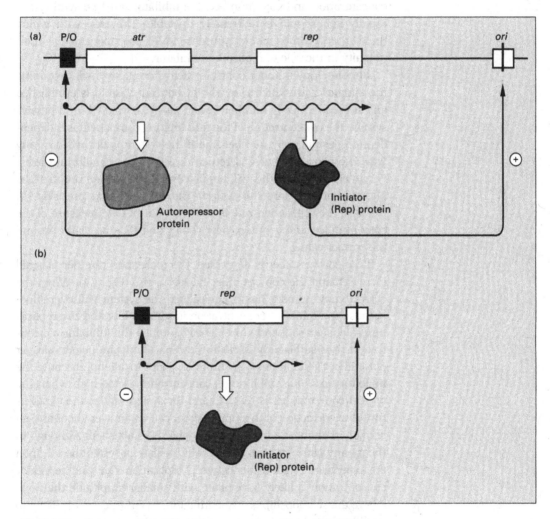

Fig. 3.3 The Autorepressor model. (a) the autorepressor (*atr*) and initiator (*rep*) genes are co-transcribed. The autorepressor regulates transcription by binding to the promoter–operator region (P/O) and maintains a constant concentration of Atr and Rep proteins. Replication is triggered by binding of Rep to the origin (*ori*). (b) In an alternative form of the Autorepressor model the Rep protein is bifunctional, acting as transcriptional repressor (which binds to p/o) and as an initiator protein which acts at the origin (*ori*).

assumed to be proportional to plasmid copy number. In reality, however, this is an impossible goal. The inhibitor is plasmid-encoded already and, when the copy number is increasing, the plasmid must exist before it can contribute its quota to the inhibitor pool. Just how far the inhibitor concentration lags behind copy number under these conditions will depend upon the rate of inhibitor synthesis. When the copy number is falling, inhibitor synthesized when the copy number was too high may persist after the situation has been corrected. To achieve the closest proportionality between inhibitor concentration and copy number, the inhibitor must be synthesized rapidly and should be extremely unstable. The ease with which these criteria can be met by antisense RNAs probably explains their ubiquity in the role of replication inhibitor.

Whether copy number control is active or passive will determine the kinetics with which a newly introduced plasmid is established and subsequent copy number deviations are corrected. In a passive system, the concentration of the rate-limiting component is independent of plasmid copy number and if the system is disturbed in any way, the copy number will return directly to its equilibrium level. For an active system this will happen only if the concentration of the plasmid-encoded inhibitor reflects the copy number precisely. In practice, the replication rate is set by an out-of-date measure of the copy number so the system over-shoots and then oscillates around the correct value.

If the copy number is depressed, the replication rate per plasmid increases and the copy number begins to rise (Fig. 3.4). When the correct copy number has been reached the system will over-shoot because the inhibitor concentration is still too low and the replication rate is consequently too high. Eventually the inhibitor concentration rises sufficiently to push the replication rate below one per plasmid per generation. Copy number begins to fall, but, because the inhibitor has finite stability, its concentration is too high when the correct copy number is reached and the system under-shoots. In its turn, the correction of this under-shoot causes an over-shoot and an oscillation about the mean copy number is established. Although in theory any active control system will oscillate, a system based upon an hyperbolic relationship between replication rate and repressor concentration is likely to oscillate with smaller amplitude than one where the relationship is a step function. In either case it is likely to be difficult to detect the phenomenon experimentally because of the difficulty of accurately determining copy number.

The difficulties of measuring the kinetics of copy number correction were overcome by Highlander and Novick (1987) who were studying derivatives of the *Staph. aureus* plasmid pT181. Plasmids were introduced by high frequency transduction and their copy

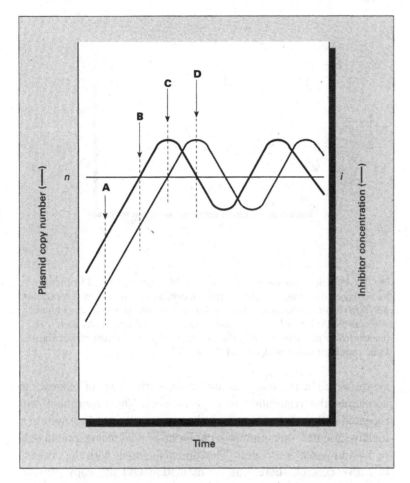

Fig. 3.4 Oscillation of active control systems. In a cell with too few plasmids, both copy number and inhibitor concentrations increase rapidly (arrow **A**). As copy number passes its mean value (n; arrow **B**) inhibitor concentration is still subnormal so copy number continues to increase. The increase in copy number stops when the inhibitor concentration reaches its mean value (i; arrow **C**). At this point copy number is high and the inhibitor concentration continues to climb, further reducing the replication rate and depressing the copy number which eventually falls below n (arrow **D**). In the absence of a damping mechanism, the system will oscillate indefinitely.

numbers were monitored during the repopulation phase. A plasmid with wild-type copy number control over-shot its copy number, reaching a value three times higher than normal before decreasing again (Fig. 3.5). This large over-shoot was shown to correlate with an absence of repressor synthesis during the first phase of repopulation. After the point of maximum copy number, repressor synthesis returned to normal and the correct copy number was restored. Presumably the amplitude of subsequent oscillations was too small to detect experimentally. The authors proposed that inhibitor

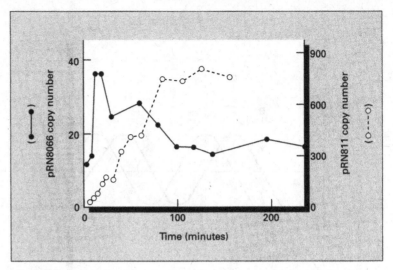

Fig. 3.5 Plasmid repopulation kinetics. Plasmids were introduced by high frequency transduction and their copy numbers monitored during repopulation. pRN8066 (with a wild-type control system) overshoots its normal copy number whereas pRN811 (in which the inhibitor-target control system has been inactivated by mutation) returns without oscillation to its accustomed value. Data from Highlander and Novick (1987).

synthesis might be repressed during the early phase of recovery to accelerate the restoration of copy number. The experiment was repeated with a plasmid in which the copy number control system is inactive and the copy number is increased 75-fold; being limited only by host-imposed constraints. This time, consistent with the absence of active control, there was no oscillation and the copy number returned monotonically to normal.

3.4 Genetic analysis of replication control

The first direct evidence that plasmids control their own replication rate was the isolation of mutants with elevated copy number (Nordström *et al.*, 1972). For more than 20 years, our understanding of the control of plasmid replication has continued to owe much to the techniques of genetic analysis. The great strength of this approach is that it permits an investigation of replication control circuits, unhindered by ignorance of their molecular nature. In this section we consider the phenotypes produced by mutations affecting various components of control systems, and how to interpret them.

3.4.1 *cop* and *rep* mutations

The control circuit in Fig. 3.2a adheres to the principles of the Inhibitor Dilution model. An unstable inhibitor (the product of the *cop* gene) is synthesized constitutively and interacts with the origin of replication (*ori*) to prevent initiation. Mutations in the *cop* gene which destroy inhibitor function cause an increase in initiation frequency and elevated copy number. Such mutations are recessive; they can be complemented in *trans* by a wild-type *cop* gene. Increased copy number could also result from a rare class of mutations which map near the origin and prevent inhibitor binding. Such mutations will be *cis*-dominant (i.e. the increased copy number is unaffected by the presence of a wild-type replicon in the same cell). Since the origin is essential for replication initiation, however, the great majority of mutations in this region will be replication-defective.

In practice, inhibitors do not act directly at the origin but block replication indirectly by preventing the expression of a Rep protein or some other component of the initiation complex (such as a primer). This is illustrated in Fig. 3.2b. As before, increased copy number results from inactivation of the inhibitor (a relatively common mutation) or its target (a much rarer class). In this case the target is the operator of the *rep* gene and operator mutations are *cis*-dominant. We would also expect to find a class of *rep* mutations which inactivate the initiator protein and result in a replication-defective phenotype.

In systems based upon autorepression of a Rep protein, similar phenotypes can be obtained. In a two-component system (Fig. 3.3a), mutations in the autorepressor (*atr*) will increase copy number and mutations in the initiator protein (*rep*) will abolish replication. When the two functions are combined in a single protein (Fig. 3.3b), it should be possible to detect a rare class of mutations which increase copy number but, paradoxically, map within the *rep* gene. These mutations destroy autorepression by preventing the binding of Rep to its operator (Trawick and Kline, 1985). Rep synthesis becomes constitutive and the copy number is increased.

3.4.2 Incompatibility studies

An additional, very powerful genetic technique used in the analysis of replication control involves the cloning of basic replicon fragments into a compatible multicopy vector and screening for the expression of incompatibility (Inc) against the parental plasmid. Incompatibility most often arises when the cloned fragment carries the replication inhibitor gene (*cop*) whose product acts in *trans* to prevent replication of the parental replicon (Fig. 3.6). The CopA repressor of R1 replication was originally identified in this way (Molin and Nord-ström, 1980).

Incompatibility testing provides a more certain way to identify primary replication inhibitors than simply screening for genes whose

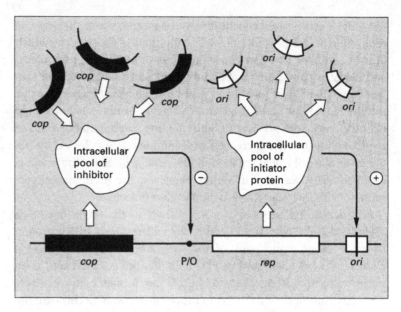

Fig. 3.6 Causes of incompatibility. The straight line represents the basic replicon of the plasmid under investigation. Fragments of this replicon inserted into compatible multicopy vectors are shown as curves. Cloned copies of *cop* supplement the concentration of inhibitor and depress replication of the parental replicon. Cloned origins titrate a *trans*-acting initiator protein away from the parental replication origin and reduce the frequency of initiation.

inactivation results in an elevation of copy number. A problem with the latter technique is that some basic replicons contain genes involved in secondary control circuits which stimulate replication under circumstances of abnormally low copy number. Examples are ColE1 *rom* and R1 *copB* whose products both decrease the frequency of replication. Rom and CopA are normally present in saturating concentration so small fluctuations in copy number (and therefore in their concentration) have no effect on replication. When these genes are cloned and their products over-expressed, their inability to further repress replication is reflected in the absence of an Inc phenotype. Inactivation of these genes by mutation mimics the conditions in a cell with extremely low copy number, leading to an increase in the plasmid replication rate and a Cop phenotype. The power of incompatibility testing is that it allows primary and secondary inhibitors to be distinguished because cloning and over-expression of primary replication inhibitors alone has an effect on replication of the parental replicon.

A less common example of incompatibility expressed by cloned replicon fragments is found in systems with a *trans*-acting Rep protein. A cloned copy of the Rep target (normally a DNA sequence overlapping or adjacent to *ori*) can act as a molecular sponge, titrating Rep protein and reducing the amount available to bind to the parent replicon (Fig. 3.6). Examples of targets identified in the way include the *incA* and *incC* iteron repeats in the P1 prophage origin (Sternberg and Austin, 1983).

3.5 The initiation of plasmid replication

Plasmid replication is controlled at the level of initiation and before embarking upon a detailed analysis of control mechanisms, it is useful to consider briefly the molecular biology of this process.

3.5.1 The role of Rep proteins

The sequence of events leading to the initiation of replication has many features in common among the majority of plasmid, bacteriophage and chromosome replicons (Fig. 3.7). It begins with the binding of at least one replicon-specific Rep protein to a set of short, repeated DNA sequences known as iterons, located close to the replication origin. At the *E. coli* chromosome origin (*oriC*), the DnaA initiator protein (analogous to the Rep protein of many plasmids and bacteriophage) binds to four 9 bp repeats (Fuller *et al.*, 1984). This

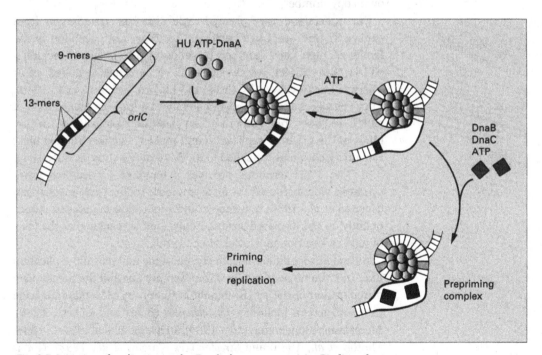

Fig. 3.7 Initiation of replication at the *E. coli* chromosome origin. Binding of DnaA to four 9 bp repeats melts an adjacent set of A : T-rich 13 bp repeats. The DnaB–DnaC helicase complex enters and further unwinds the template, revealing sites for priming and initiation of replication. Reproduced with permission from Bramhill and Kornberg (1988a).

promotes melting of an adjacent series of thermodynamically unstable 13 bp A : T-rich repeats, allowing access of the DnaBC helicase complex (Bramhill and Kornberg, 1988b; Kowalski and Eddy, 1989; Gille and Messer, 1991; Hsu *et al.*, 1994). The helicase unwinds the region surrounding the origin, revealing priming sites where DnaG protein synthesizes RNA primers for extension by polymerase III holoenzyme. The timing of initiation at the chromosome origin is influenced by multiple factors in addition to the availablility of DnaA. These include transcription, methylation status and membrane attachment of the origin region, and the concentration of the IciA protein which binds to the 13 bp repeats *in vitro* and prevents DnaA-mediated melting (Hwang and Kornberg, 1990; Thony *et al.*, 1991).

The organization of many non-chromosome origins (including F, P1, R1 and λ) resembles *oriC* and the sequence of events leading to initiation is triggered by the binding of replicon-specific proteins (reviewed by Zyskind and Smith, 1986; Bramhill and Kornberg, 1988a). DnaA often plays an accessory role at these origins but may not be essential (Abeles *et al.*, 1990; Bernander *et al.*, 1991). By encoding their own initiator proteins, these elements are able to replicate independently of the chromosome and thus define their own copy number.

Our understanding of plasmid replication has come largely from studies in Gram-negative bacteria. An important exception is the family of high copy number plasmids of *Staph. aureus* including pT181 and pC194 which replicate *via* a single-stranded DNA intermediate (reviewed by Gruss and Ehrlich, 1989; Novick, 1989). They reveal a quite different role for the Rep protein. Replication of pT181 requires the synthesis of an initiator protein (RepC) which binds to the origin of replication and makes a single-stranded nick, remaining covalently attached to the 5' terminus thus produced (Fig. 3.8). The 3'OH terminus provides a template for leading-strand synthesis which proceeds by an asymmetric rolling circle mechanism (Koepsel *et al.*, 1986). Lagging-strand replication is initiated subsequently on the displaced strand. Replication is regulated at the level of RepC transcription (Novick *et al.*, 1989).

Although very different from chromosome and plasmid replication in *E. coli*, the mode of action of the Rep protein and the mechanism of replication of the pT181 family is shared by plasmids in a wide range of genera including *Lactobacillus* (Bates and Gilbert, 1989), *Mycoplasma* (Bergemann *et al.*, 1989), *Streptococcus* (Lacks *et al.*, 1986; Minton *et al.*, 1988) and *Streptomyces* (Zaman *et al.*, 1993). Nor is rolling circle replication restricted to the Gram-positives; it is also found among the single-stranded coliphages (Dotto *et al.*, 1984) and plasmids of the Gram-negative bacterium *Helicobacter pylori* (del Solar *et al.*, 1993). Rolling circle plasmids thus represent a group of

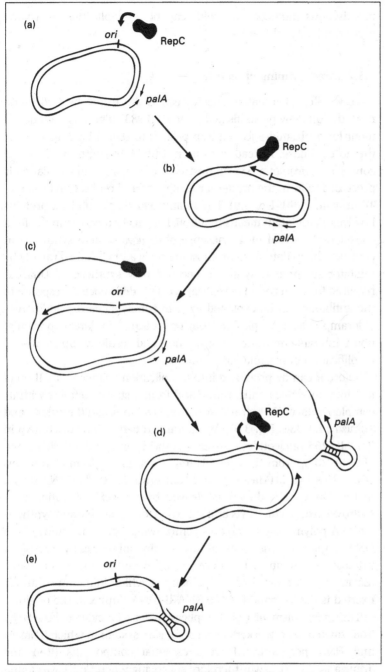

Fig. 3.8 Initiation of pT181 replication. (a) The RepC initiator protein makes a single strand break at the origin (*ori*), remaining attached to the 5′ terminus of the nick. (b & c) The exposed 3′-OH serves as a primer for leading strand DNA synthesis by a rolling circle mechanism. (d & e) RNA polymerase primes lagging strand synthesis at the *palA* locus (Gruss *et al.*, 1987) where a hairpin loop forms in the displaced single strand. If lagging strand synthesis initiates after the completion of leading strand replication, single strand circular intermediates may accumulate. Their detection is diagnostic of this mode of replication.

promiscuous multicopy plasmids capable of replication in a wide variety of bacterial species.

3.5.2 Direct priming of replication

A very different initiation mechanism is illustrated by ColE1 and related multicopy plasmids (Selzer *et al.*, 1983). There is no requirement for a plasmid-encoded Rep protein to reveal binding sites for the host primase. Instead, a transcript (RNA II) synthesized from a constitutive promoter 555 bp upstream of the origin of replication is processed to form the primer for leading strand replication (Itoh and Tomizawa, 1980; Fig. 3.9). The relative stability of all host proteins involved in initiation means that ColE1 replication can continue for a considerable period in the absence of *de novo* protein synthesis. In contrast, the initiation of chromosome replication is halted rapidly by inhibitors of protein synthesis because new synthesis of DnaA is required for each round of replication. This difference in response to the antibiotic can be exploited experimentally since the addition of chloramphenicol to plasmid-bearing cultures in late exponential phase inhibits chromosome replication and results in up to 50-fold amplification of plasmid copy number.

Before it can be processed into a replication primer, RNA II must fold into an active configuration and form a stable complex with its complementary DNA strand at the replication origin (Masukata and Tomizawa, 1984; Fig. 3.9). An interaction between a G-rich loop in RNA II, 265 nucleotides upstream of the 3' end, and a C-rich patch of DNA 20 bp from the origin is implicated in the formation of this DNA–RNA hybrid (Masukata and Tomizawa, 1990). The RNA strand in this DNA–RNA duplex is cleaved by RNaseH, liberating a 3' hydroxyl group which acts as a substrate for leading-strand synthesis by DNA polymerase I (Itoh and Tomizawa, 1980). Unwinding of the DNA duplex downstream of the origin subsequently reveals a primosome assembly site where lagging-strand synthesis is initiated (Zavitz and Marians, 1991). The point at which initiation control is exerted is the formation of the RNA II–DNA duplex at the origin.

The mechanism of ColE1 replication was elucidated largely by Tomizawa and co-workers using an *in vitro* system in which RNaseH and DNA polymerase I were essential components (Itoh and Tomizana, 1978). It was therefore surprising when ColE1 was found to replicate not only in RNaseH-deficient mutants (Naito *et al.*, 1984) but also in mutants deficient in both RNaseH and DNA polymerase I (Kogoma, 1984). In common with the initiation pathway established *in vitro*, this alternative mode of replication *in vivo* requires RNA II to hybridize to DNA at the replication origin. It is thought that in the absence of RNaseH and PolI, the RNA II–DNA hybrid extends

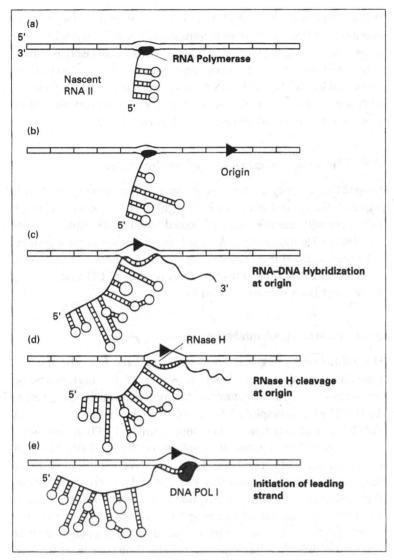

Fig. 3.9 Initiation of ColE1 replication. (a) Transcription of the preprimer RNA (RNA II) is initiated 555 bp upstream of the origin by RNA polymerase. (b & c) As the transcript is elongated it folds into a configuration which assists the formation of a stable RNA–DNA hybrid at the origin of replication. (d) RNaseH cleaves the RNA strand in the RNA–DNA duplex at the origin. (e) The 3′OH terminus of the RNA serves as a primer for the initiation of leading strand replication by DNA polymerase I. Reproduced with permission from Polisky (1988).

beyond the normal origin of replication, displacing the non-transcribed strand (Dasgupta *et al.*, 1987; Masukata *et al.*, 1987). A minimum length of 40 bp must be displaced to admit the helicase which promotes further unwinding of the duplex and subsequent formation of a replisome (Kornberg, 1982). Lagging-strand synthesis is then initiated without the involvement of DNA polymerase I.

When polymerase I is present, a third initiation mechanism may operate with RNA II transcripts terminated beyond the origin acting as primers for leading-strand synthesis. Because RNaseH is inhibitory to the polymerase I independent pathway, it is unlikely to be important in wild-type cells but it does provide an interesting parallel with the mechanism of initiation of many Rep protein dependent replicons where strand separation is the crucial step.

3.6 The control of plasmid replication

Despite the diversity of mechanisms by which plasmid replication is initiated, the control circuits which regulate this process are largely consistent with the principles embodied in either the Autorepressor or Inhibitor Dilution model. A useful way to classify control systems is by the nature of their primary replication inhibitor which may be a protein (λ-dv), a small antisense RNA (ColE1, pT181 and R1) or a set of short DNA repeats (P1 and F).

3.6.1 Antisense RNA inhibitors

One of the most unexpected discoveries to come from the study of plasmid replication control was the ubiquity of short, antisense transcripts in the role of primary inhibitor (reviewed by Eguchi *et al.*, 1991). They are employed in *E. coli* both by high copy number ColE1-like plasmids and by low copy number IncFII replicons. In *Staph. aureus* they regulate the replication of the pT181 family of multicopy plasmids. The basic designs of these control circuits are very similar despite the diversity of plasmid size, copy number and the mechanisms by which replication is initiated. In each case, the short antisense inhibitor binds near the 5' end of a longer transcript which plays an essential part in the train of events leading to replication. Inhibitor binding prevents the transcript from fulfilling its role in this process. In ColE1, the inhibitor blocks the formation of an RNA II–DNA hybrid at the replication origin. In R1 and pT181 the inhibitor binds to the mRNA of an essential Rep protein, preventing translation or causing premature termination of the message.

3.6.2 ColE1-like plasmids

Probably the most-studied antisense transcript of all is the 108 nucleotide RNA I inhibitor of ColE1 replication (Fig. 3.10; for reviews see Polisky, 1988 and Cesareni *et al.*, 1991). Elucidation of the mechanism of ColE1 replication control has involved an elegant and powerful alliance of biochemistry and genetics. The majority of studies employ an *in vitro* system in which the initiation of

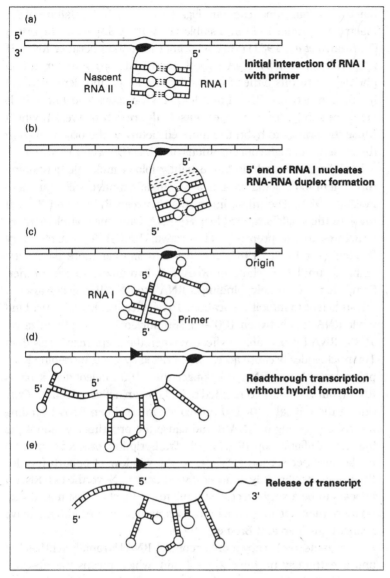

Fig. 3.10 Inhibition of ColE1 replication by RNA I. (a) RNA II transcription initiates at a promoter 555 bp upstream of the replication origin. (b) RNA I interacts with the growing RNA II transcript. (c) This interaction causes RNA II to adopt a structure which is unable to form a stable DNA–RNA hybrid at the origin of replication. (d & e) In the absence of hybrid formation at the origin there is no RNaseH cleavage of the preprimer and no template is available for DNA polymerase I. When transcription terminates downstream of the origin, the inactive RNA II is released. Reproduced with permission from Polisky (1988).

replication requires template DNA, DNA-dependent RNA polymerase, RNaseH and DNA polymerase I (Itoh and Tomizawa, 1978).

RNA I is complementary to the 5′ end of the RNAII preprimer and binding between these RNAs causes the preprimer to fold into an inactive configuration (compare Figs 3.9 and 3.10). The failure of the inactive preprimer to form a stable DNA–RNA duplex at the origin (Tomizawa *et al.*, 1981) prevents primer formation because RNaseH is specific for a DNA–RNA hybrid. Increased copy number (a Cop phenotype) results either from mutations which interfere with the interaction between RNA I and RNA II (Tomizawa and Itoh, 1981) or, more subtly, from changes outside the region of overlap which cause the primer to hybridize more efficiently to the origin, even in the presence of high concentrations of RNA I (Fitzwater *et al.*, 1992).

RNA I exists in solution as a tightly folded molecule containing three stem-loop domains and a 5′ single-stranded tail (Tam and Polisky, 1983). The initial interaction between RNA I and RNA II involves the single-stranded loops of RNA I and the complementary structures in the preprimer (Tomizawa, 1990a). This exploratory 'kissing' (Tomizawa, 1984) develops into an ever more passionate embrace until the whole of RNA I is hybridized to the primer. Somewhat surprisingly, binding of RNA I to RNA II is necessary but not sufficient to inhibit replication. To be effective, RNA I must bind while RNA II is between 100 and 360 nucleotides long (Tomizawa, 1986). RNA I is probably ineffective when the preprimer is less than 100 nucleotides because access is restricted by the transcribing RNA polymerase. When RNA II is longer than 360 nucleotides, it binds RNA I, but is already committed to fold into an active structure. Only during the critical 100–360 nucleotide interval can RNA I binding modify the folding of RNA II and cause the preprimer to adopt the inactive configuration (Fig. 3.10). Transcription proceeds at ≈ 60 nucleotides per second at 37 °C so the window of opportunity for RNA I exists for only four or five seconds. Nevertheless RNA I appears to be a very successful inhibitor; it has been estimated that no more than one in 20 origin transcripts is processed into an active primer (Lin-Chao and Bremer, 1987).

As befits its role in copy number control, RNA I is rapidly synthesized and is extremely unstable ($t_{1/2} = 2$ min) which means that its concentration provides an accurate estimate of plasmid copy number. Turnover of RNA I begins with the docking of its 5′ single-stranded tail by RNAse E, giving rise to an unstable 103 nucleotide form known as RNA I_{-5} or RNA I. The precise fate of this cleaved form is unclear but observations that plasmid copy number is reduced and RNA I accumulates in a host lacking PcnB (a poly-A polymerase), has led to the proposal that polyadenylation of may be important during subsequent degradation (Lin-Chao and Cohen, 1991; He *et al.*, 1993).

3.6.3 Rom: a matchmaker protein

An additional component of the ColE1 copy number control system was identified after it was observed that deletions on the far side of the origin from the RNA II coding sequence cause an increase in copy number (Twigg and Sherratt, 1980). This region contains the *rom* gene which encodes a *trans*-acting inhibitor of replication. Rom (known originally as Rop because of an incorrect hypothesis about its function; Cesareni *et al.*, 1982) increases the rate at which RNA I binds to the preprimer transcript (Tomizawa and Som, 1984). A detailed analysis of the effect of Rom on the kinetics of the interaction between RNA I and RNA II has been made by Tomizawa (1990b). Because only preprimer transcripts between 100 and 360 nucleotides are susceptible to RNA I, the rate at which RNA I binds to its target is critical to its efficiency as an inhibitor. RNA footprinting studies show that the 63 amino acid Rom polypeptide protects the double-stranded stems of the two complementary transcripts, but makes the single-stranded loops more accessible to nuclease attack (Helmer-Citterich *et al.*, 1988). Rom seems to act as a 'matchmaker' protein, stabilizing the initial kissing complex (Fig. 3.10a & b) and ensuring the appropriate positioning of the complementary loops.

Rom is not an essential component of the ColE1 control circuit; if the *rom* gene is deleted, copy number is raised but still regulated. If *rom* is cloned on a compatible multicopy plasmid, it has no effect upon the replication of a co-resident ColE1 derivative. The absence of an incompatibility reaction demonstrates that Rom is not the primary inhibitor of ColE1 replication because it is already exerting its maximum effect at the wild-type concentration.

3.6.4 pT181: transcriptional attenuation

The *copA* region of pT181 encodes two short transcripts RNA I and RNA II (beware confusion with the key players in the ColE1 story!). They are synthesized from the same promoter and are complementary to the 5′ end of the messenger RNA for the RepC initiator protein (Fig. 3.11). When an inhibitor transcript forms an RNA duplex with the RepC message, the folding of the message is altered so that a rho-independent terminator forms and transcription stops immediately upstream of the start codon. In the absence of inhibitor binding, the RepC message folds into a secondary structure which preempts formation of the terminator stem-loop. The system is analogous to attenuators that regulate amino acid biosynthesis genes in *E. coli* and has been dubbed a 'countertranscript-driven transcriptional attenuator' (Novick *et al.*, 1989). Under normal conditions, only 3% of transcripts from the *repC* promotor escape premature termination. The importance RNA secondary structure in the control of pT181 replication is illustrated by the effect of a mutation which, by lowering the free energy of the terminator stem from

−13.9 kcal mol^{-1} to −6.0 kcal mol^{-1}, increases the copy number 25 fold (Carleton *et al.*, 1984).

Genetic analysis of pT181 reveals two regions which, when cloned, express incompatibility against the parent replicon (Highlander and Novick, 1990). Inc3A is associated with *copA* which encodes the inhibitor transcripts and their target. This is precisely analogous to the Inc phenotype associated the RNA I–RNA II overlap region of ColE1. The second (Inc3B) is associated with the leading-strand replication origin. If additional origins cloned on a multicopy plasmid are introduced into the cell, there is competition for the limited amount of *trans*-acting RepC protein. The share of RepC which binds to the origin of pT181 is thus reduced and its replication rate and copy number decreased. The ability of the

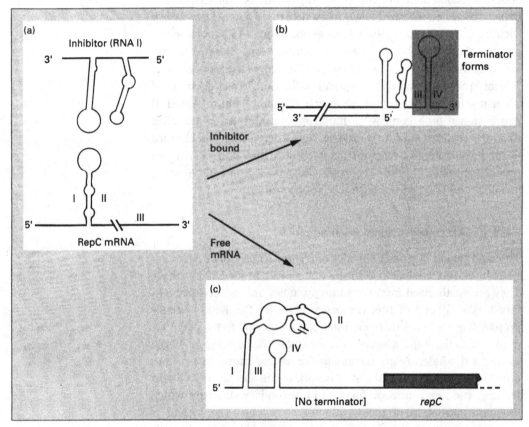

Fig. 3.11 Control of pT181 replication. The RNA I inhibitor causes premature termination of RepC mRNA. (a) The initial interaction between RepC mRNA and RNA I is through complementary single strand loops (very similar to ColE1 RNA I–RNA II, Fig. 3.10). (b) Inhibitor binding allows base pairing between regions III and IV a terminator forms before the start of *repC*. (c) In the absence of the inhibitor, regions I and III associate; no terminator is formed and transcription continues into *repC*. Reproduced with permission from Novick *et al.* (1989).

pT181 origin to compete for RepC is enhanced by the *cmp* locus (Gennaro and Novick, 1988). Its precise mechanism of action is unclear but it may bind a host factor which increases the efficiency of the interaction between RepC and the origin.

3.6.5 Plasmid R1: antisense control of a low copy number plasmid

Replication control by antisense transcripts is not restricted to high copy number plasmids. R1 is a low copy number self-transmissable plasmid belonging to the IncFII group (reviewed by Nordström *et al.*, 1984). Control is at the level of translation of an initiator (RepA) protein. Mutations causing elevated copy number map to the *copA* and *copB* regions of the basic replicon (Fig. 3.12) and identify two independent inhibitors of replication (Stougaard *et al.*, 1981; Riise *et al.*, 1982).

copB encodes a polypeptide which inhibits transcription from ^{P}*repA*; one of the two promoters for transcription of *repA*. Despite its repressor activity, CopB is not a component of the primary control circuit. Further analysis revealed that the introduction of extra copies of *copB* on a compatible, multicopy plasmid had no effect upon R1 replication (i.e. it did not display an Inc phenotype). It appears

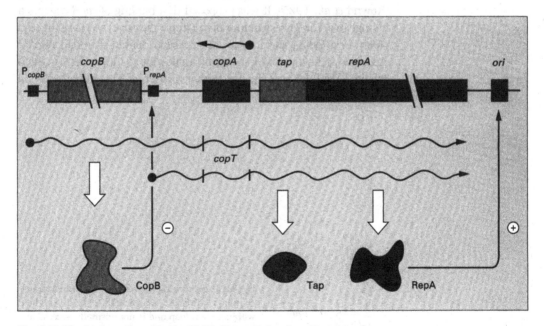

Fig. 3.12 The basic replicon of plasmid R1. Transcription from P$_{repA}$ is repressed by CopB unless copy number is abnormally low. Under normal conditions *copB*, *tap* and *repA* are co-transcribed from P$_{copB}$. The CopA transcript binds to CopT mRNA and inhibits translation of *tap* and *repA*.

that $^{\text{P}}repA$ is fully-repressed for most of the time and extra CopB has no effect. Under normal conditions, *repA* is co-transcribed with *copB* from $^{\text{P}}copB$. Only in cells with very low copy number and a low concentration of CopB (e.g. immediately after conjugation) is transcription from $^{\text{P}}repA$ induced.

copA produces a small, unstable transcript which is complementary to the 5′ untranslated leader of RepA mRNA, about 80 nucleotides upstream of the start of RepA translation. Consistent with its role as the primary inhibitor of replication, a multicopy plasmid carrying *copA* is incompatible with wild-type R1. The transcript interacts with *copT*, the complementary region of RepA mRNA (Persson *et al.*, 1988), inhibiting translation of the message. Despite much study and speculation, the mechanism by which CopA inhibits RepA translation remained unclear for many years. Following the discovery that the CopA–RepA mRNA duplex is a substrate for RNAase III (Blomberg *et al.*, 1990), it was suggested that rapid degradation of the message triggered by RNase III cleavage might be important. In fact cleavage reduces message stability by only two- or three-fold which is insufficient to account for the effect of CopA on RepA expression. What is more, full duplex formation is unnecessary for inhibition of RepA synthesis, a 'kissing' interaction between inhibitor and target is sufficient (Wagner *et al.*, 1992). An alternative hypothesis followed computer simulations of RepA mRNA folding (Rownd *et al.*, 1985). It was proposed that binding of *copA* triggers a change in secondary structure of the RepA message, sequestering the ribosome binding site into an inaccessible double-stranded region. Acting in this way, CopA would be analogous to ColE1 RNA I which modifies the folding of the pre-primer transcript. This hypotheses fell

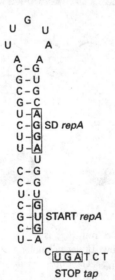

Fig. 3.13 The mechanism of *tap–repA* translational coupling. The Shine–Delgarno (SD) site of *repA* lies within a stem-loop and is not normally accessible for ribosome binding. The *tap* open reading frame extends beyond the start of *repA* and translation of *tap* mRNA 'irons-out' the stem loop, allowing ribosomes to bind and translate *repA*.

out of favour when the computer predictions were not supported by a study of RNA structure *in vitro* (Öhman and Wagner, 1989).

More recently it has been shown that CopA acts indirectly by inhibiting translation of a 24 amino acid leader polypeptide (*tap*; translational *a*ctivator *p*olypeptide). CopA binds just upstream of the *tap* ribosome binding site and is proposed to prevent access by the translation machinery. Translation of *tap* is coupled to *repA* translation (Blomberg *et al.*, 1992). Ribosomes cannot bind directly to the *repA* ribosome binding site as both this and the GUG start codon are sequestered into a stable stem-loop structure. Access to the ribosome binding site is possible only when the stem-loop is ironed-out by the passage of ribosomes translating *tap* (Fig. 3.13). Thus *repA* is translated only in the absence of CopA. The translational coupling hypothesis has gained support from the demonstration that disruption of the stem-loop by mutation permits *tap*-independent *repA* expression (Blomberg *et al.*, 1994).

3.6.6 λ-*dv*: regulation by a *trans*-acting protein

Early models of plasmid copy number control assumed that the role of primary inhibitor would be filled by *trans*-acting proteins. In fact, protein inhibitors are rare among replication control circuits. One basic replicon where the primary inhibitor is a protein is λ-*dv* (Fig. 3.14); an autonomously replicating fragment of the bacteriophage λ genome containing the O and P protein genes which are required for bacteriophage replication (Matsubara, 1976; Murotsu and Matsubara, 1980). It is also worthy of note as an excellent realisation of the principles outlined in the Autorepressor model (Sompayrac and Maaloe, 1973; Fig. 3.3a).

Replication of λ-*dv* requires the synthesis of the replicon-specific O and P proteins which bind to the replication origin and trigger initiation. The λ O protein binds first, followed by a complex between the DnaB helicase and the λ P protein. Release of the P protein activates the helicase which unwinds the origin region, allowing primer synthesis by DnaG and subsequent extension by polymerase III holoenzyme. The O and P proteins are transcribed from the P_R promoter as part of a polycistronic mRNA which also contains the coding region for the Cro protein (known originally as Tof). Cro is the primary inhibitor of replication; by binding to the O_R operator it acts as an autorepressor of transcription from P_R. This control circuit ensures that the cellular concentration of O and P, and therefore the total replication frequency per generation, are constant. In such a system the replication frequency per plasmid is inversely proportional to copy number, thus satisfying the basic criterion of replication control.

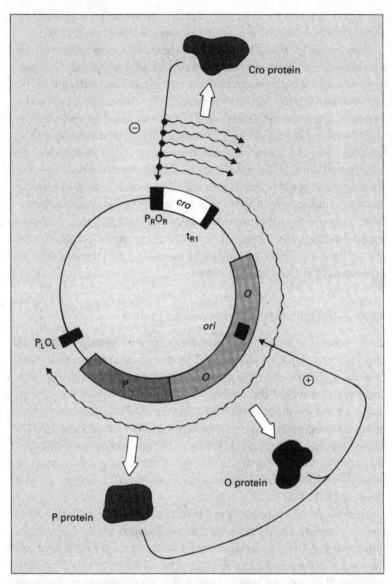

Fig. 3.14 Replication of λ-*dv* plasmids is triggered by binding of the O and P proteins to the origin. Genes *cro*, *O* and *P* constitute an operon transcribed from p_R. 80% of p_R transcripts terminate at t_{R1}. By binding at o_R, Cro protein autoregulates transcription from p_R and maintains a constant concentration of Cro, O and P. After Murotsu and Matsubara (1980).

3.6.7 The P1 prophage: titration and handcuffing

For many basic replicons the nature of the inhibitor is clear; a short, antisense transcript for plasmids in the ColE1, pT181 and R1 families or a protein for λ-*dv*. In some cases (including F and the P1 prophage) the identity of the inhibitor is not immediately obvious but genetic and biochemical studies suggest that the replication rate is controlled by a series of short DNA repeats (known as iterons) which bind the Rep protein. The use of DNA repeats as *trans*-acting inhibitors of replication appears to be a common form of control in plasmids of both eukaryotes and prokaryotes (Nordström, 1990).

P1 is a bacteriophage whose prophage exists as a plasmid. Prophage replication is limited not simply by the availability of the RepA initiator protein but by whether it is able to bind to the replication origin. Genetic analysis of the R replicon of P1 (Sternberg and Austin, 1983) identified two regions (*incA* and *incC*; Fig. 3.15a) which, when cloned, exert incompatibility against their parent replicon and are therefore candidates for the roles of inhibitor or replication origin. Each consists of a series of 19 bp iterons which bind RepA (Abeles, 1986). Although similar in sequence and organization, the two sets of repeats have quite distinct functions. *incC* contains the replication origin and RepA binding triggers initiation. RepA also binds to *incA* (the inhibitor repeats) and this effectively inactivates the initiator by titrating it away from the replication origin.

In the absence of *incA* repeats, RepA and *incC* maintain effective replication control despite a 10-fold increase in copy number. The *repA* promoter-operator is located within *incC* and transcription is autoregulated (Chattoraj *et al.*, 1985), maintaining RepA at an average concentration of 20 dimers per cell (Swack *et al.*, 1987). This simple control circuit is entirely consistent with the principles of the Autorepressor model. When *incA* is introduced (either in *cis* or in *trans*) the plasmid copy number is reduced; an effect proposed initially to be due to titration of the Rep protein away from the origin. Despite the plausible simplicity of a combined titration–autorepressor model, this cannot be the whole story. Any model which simultaneously invokes a constant concentration of Rep protein maintained by autoregulation and titration of the protein by the control repeats, cannot work. Titrated Rep would be unable to bind to the operator and would be replaced by an elevated rate of transcription. In such a system control over replication is lost unless *incA*-bound RepA can still repress transcription from the *repA* promoter.

A solution to the titration–autorepressor paradox was provided by Chattoraj *et al.* (1988) who observed by electron microscopy that RepA can bind simultaneously to *incA* and *incC*, pairing the sites and causing the intervening DNA to loop (Fig. 3.15b). If such a paired structure is unable to initiate replication, the observation can account for the failure of excess RepA to increase P1 copy number in the presence of *incA* (Pal and Chattoraj, 1988).

In an extension of the looping model, Abeles and Austin (1991) argue that although *cis*-looping could set the maximum level of origin activity, pairing in *trans* (or 'handcuffing') must be invoked to achieve true copy number regulation (Fig. 3.15c). Their proposal is

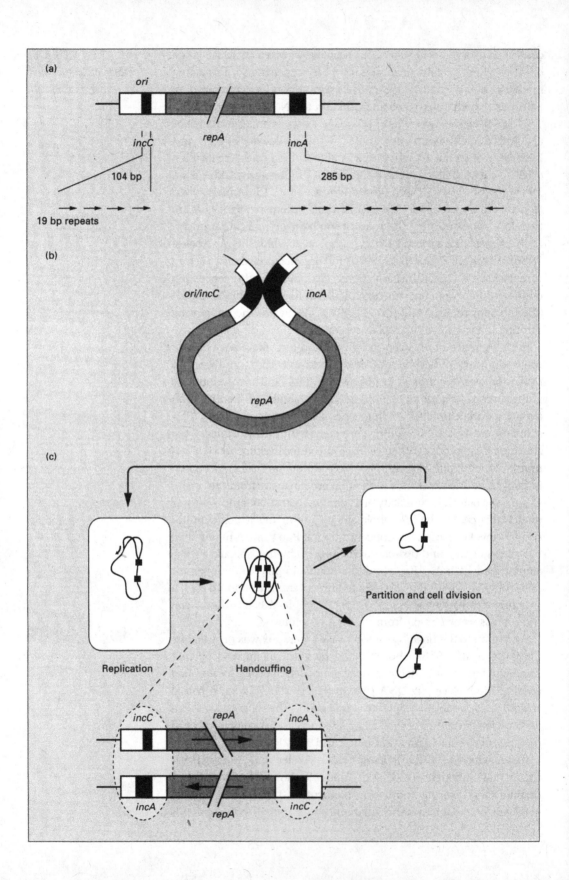

lent credibility by the high frequency of *trans*-paired structures seen in the electron microscope. They propose that after replication the daughter molecules pair in an antiparallel fashion, blocking further initiation, even in the presence of excess RepA. Eventually plasmid partition and cell division separate the plasmids which are then free to replicate. Replication during a defined phase of the cell cycle, which is predicted by this model, has been reported by Keasling *et al.*, (1992) although the evidence is not unequivocal (Nordström and Austin, 1993). It also suggests that replication should shut down completely when the correct copy number has been achieved. Tsutsui and Matsubara (1981) have demonstrated replication shutdown for plasmid F which has a very similar arrangement of origin and control repeats to P1.

Even in the handcuffing model we may not have a complete list of the *dramatis personae* in P1 replication. It does not explain why, even in the absence of *incA*, five-fold over-production of RepA inhibits replication and 40-fold over-production stops replication altogether (Muraiso *et al.*, 1990). The identity of the inhibitor is unknown but the involvement of an unstable fragment of the Rep protein has been suggested.

The concept of control by handcuffing was proposed independently to explain replication control in plasmids R6K (McEachern *et al.*, 1989) and RK2 (Kittell and Helinski, 1991; Durland and Helinski, 1990). These plasmids differ from P1 by the absence of *incA*-like control repeats and direct origin–origin contacts are invoked.

3.7 The two-tier organization of control circuits

The apparent complexity of many replication control systems reflects the presence of two control circuits. Although a single negative feedback loop is in principle sufficient for copy number regulation, secondary circuits may reduce the response time of the system to large deviations in copy number. This helps to maintain a low copy number variance in the population, which is important for stable plasmid maintenance (see chapter 3). In both ColE1 and R1, the secondary control circuit involves *trans*-acting repressor proteins (Rom and CopB, respectively). Deletion or mutation of the repressor gene leads to an increase in plasmid copy number but, unlike

Fig. 3.15 (*Facing page*) Control of P1 replication. (a) Organization of the P1 basic replicon. *repA* encodes an initiator protein which binds a series of 19 bp repeat sequences within *incC* and *incA*. From Pal *et al.* (1986). (b) Electron microscopy provides evidence for RepA-mediated pairing of *incA* and *incC*. (c) The handcuffing model. The products of replication associate by *trans* pairing of *incA* and *incC*. The paired structure is unable to initiate replication until the structure has been disrupted by active partition at cell division.

primary inhibitors, over-production has no effect upon plasmid replication.

3.7.1 CopB: a secondary inhibitor of R1 replication

CopB is a secondary inhibitor of R1 replication. Under most circumstances CopB represses transcription from the $^P repA$ promoter (Light and Molin, 1982; Riise *et al.*, 1982) and *repA* is transcribed by readthrough from $^P copB$ (see Fig. 3.12). The replication rate is modulated by the interaction of CopA with the complementary region of the *repA* message. Small copy number fluctuations during vegetative replication are unlikely to cause instability because R1 is partitioned actively at cell division. Immediately after conjugation, however, the recipient cell contains only one plasmid and if cell division occurs before replication, a plasmid-free daughter will be produced. In this situation, rapid replication of the plasmid is obviously desirable (Womble and Rownd, 1986a) and because the recipient contains no CopB protein, transcription of *repA* initiates at both $^P repA$ and $^P copB$. This increases the supply of RepA and a burst of rapid replication results. As the plasmid copy number rises, CopB is produced and the $^P repA$ promoter is again fully repressed.

3.7.2 The role of ColE1 Rom

ColE1 replication control incorporates a secondary repressor (Rom) which, like R1 CopB, boosts the replication rate in low copy number cells. Although functionally analogous, the molecular mechanisms of CopB and Rom action are very different. Whereas CopB inhibits transcription of the R1 *rep* gene directly, Rom assists the interaction of the RNA I inhibitor with its target, thereby increasing its effectiveness (Fig. 3.10). Like R1, ColE1 can transfer horizontally in bacterial populations although the majority of its transfer functions must be supplied by a co-resident, self-transmissible plasmid. ColE1 is distributed randomly between daughters at cell division and there is a high risk of plasmid loss if the recipient cell divides soon after plasmid transfer. Rapid post-conjugal replication is ensured by the initial low concentration of both RNA I and Rom in the recipient. Most preprimer transcripts fold into the active configuration because of the low concentration of RNA I and because RNA I binds inefficiently to the pre-primer in the absence of Rom.

3.8 Quantitative modelling of plasmid replication control

The relative simplicity of replication control systems has encouraged attempts to construct computer simulations which mimic their behaviour. The performance of each model is judged by its ability to maintain a stable copy number and by its response to changes in parameters such as host cell growth rate or the kinetics of interaction between a replication inhibitor and its target. Model building can provide a useful test of our understanding of the key interactions in a specific system but the results must be treated with caution. Copy number control can be achieved in many different ways and it follows that it may be possible to construct a plausible model which reproduces the characteristics of a particular biological system but which is based upon erroneous assumptions. This is illustrated by models of the mini-F control system based upon the Autorepressor hypothesis. The models were consistent with the available data (Trawick and Kline, 1985; Womble and Rownd, 1987) but contained as a central concept a transcriptional repressor which was processed slowly into a replication initiator protein. This metamorphosis has never received experimental support and a revised model, arising from work with the P1 prophage, in which the same protein fulfills both roles, is now in the ascendancy.

In view of the extensive genetic and biochemical analysis of bacteriophage λ, the λ-*dv* plasmid is an attractive prospect for the construction of a rigorous replication control model. A quantitative analysis (Lee and Bailey, 1984a) successfully predicted the plasmid copy number and repressor concentration throughout the cell cycle. The same model has been used to simulate the phenotypes of mutations affecting the Cro repressor or the transcription of initiator protein genes (Lee and Bailey, 1984b), and its predictions are generally consistent with experimental observations. The basic framework of the λ-*dv* model has been developed further (Womble and Rownd, 1986b) and has been used as the basis of a model of replication control for the IncFII plasmid NR1 (Womble and Rownd, 1986a). The latter has been criticized, however, for failing to recognize the non-equilibrium nature of the interaction between the repressor and its target (Persson *et al.*, 1990).

Ultimately, mathematical modelling of the processes which underpin plasmid replication may facilitate the design of vectors with properties customized to suit the user. The majority of vectors in current use contain high copy number ColE1-like replicons. Experimental estimates of RNA I, RNA II and Rom concentrations *in vivo* have been used by Brenner and Tomizawa (1991) to investigate the response of the ColE1 copy number control system to changes in inhibitor and target concentrations. They found that because the majority of pre-primer transcripts are not used for replication, the system is relatively insensitive to changes in the rate of RNA II synthesis. In contrast, plasmid copy number is strongly dependent on the concentration of RNA I.

Ataai and Shuler (1986) attempted to make *a priori* predictions of copy number for ColE1-like plasmids based soley upon independently determined parameters. The authors demonstrate the use of their model to distinguish among different possible modes of replication. Plasmids are (i) selected at random for replication (ii) replicate exactly once per generation or (iii) only one plasmid in each cell is able to replicate (the master copy hypothesis). Only the random replication model which is favoured by independent experimental evidence; (Bazoral and Helinski, 1970; Summers *et al.*, 1993) produces an outcome which is consistent with observed plasmid copy number.

4: Plasmid Inheritance

4.1 The nature of plasmid instability

It is essential at the outset to clarify what exactly is meant by 'plasmid instability'. Segregational and structural instability are twin spectres which haunt the plasmid worker. The former is the consequence of a daughter receiving no plasmid DNA at cell division and giving rise to a clone of plasmid-free descendants. In this chapter we analyse the causes of segregational instability and the strategies adopted by natural high and low copy number plasmids to ensure stable maintenance (for a general review see Nordström and Austin, 1989). Structural instability involves the rearrangement or loss of plasmid DNA sequences, typically associated with transposition or recombination. It will not be addressed further in this chapter but its importance as a source of genetic variation in the plasmid gene pool is apparent in discussions of plasmid evolution in Chapter 2.

4.1.1 Segregational instability

Recent years have seen much effort expended in the study of segregational instability. A major stimulus for this work has been the development of synthetic cloning vectors which lack the stability of naturally occurring plasmids. Although rarely more than a minor irritation to the research worker, instability can spell disaster for the biotechnologist seeking efficient expression of a plasmid-borne gene in a large fermentor.

For a plasmid to be stably inherited, a number of conditions must be fulfilled. First, each plasmid must replicate, on average, once each generation and copy number deviations must be corrected. In addition, the products of replication must be distributed to both daughters when the cell divides. In most cases, instability results from a failure of distribution at cell division and it is on this process that we will concentrate our attention.

4.2 Plasmid distribution at cell division

4.2.1 Active partition

Anucleate cells arise very rarely in bacterial culture. This implies that there exists some machinery for the distribution or partition of chromosomal DNA to both daughters. It has long been assumed that a similar process operates for low copy number plasmids (Jacob *et al.*, 1963). Two mechanisms of active partitioning have been proposed (Nordström *et al.*, 1980a; Fig. 4.1a). Equipartition involves the distribution of exactly half the plasmid molecules to each cell. This is analogous to mitosis during somatic cell division in eukaryotes. Pair-site partitioning envisages that only one pair of plasmids is partitioned actively and the remainder are distributed randomly. Either of these systems operating in conjunction with an efficient copy number control system will confer a high degree of plasmid stability.

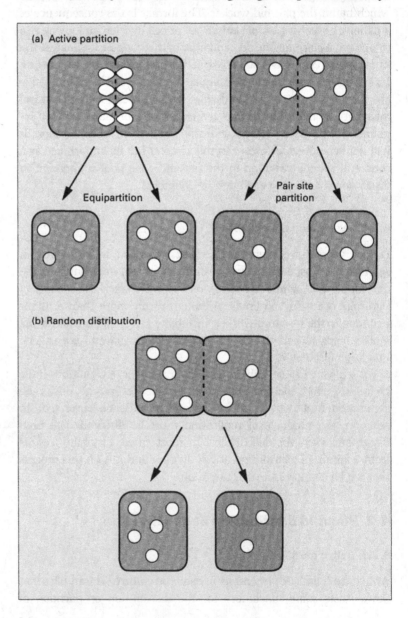

Fig. 4.1 Plasmid distribution at cell division. (a) Active partition. (b) Random distribution.

4.2.2 Random distribution

A radical alternative to active partition is the random distribution of plasmids between daughter cells (Fig. 4.1b). This is unsuitable for low copy number plasmids such as F, but at a sufficiently high copy number, the segregation frequency (the frequency at which plasmid-free cells arise) may be low enough that the plasmid appears stable. For randomly distributed plasmids this frequency can be readily calculated. The probability that a plasmid molecule will *not* enter a specific daughter cell is 0.5. If there are n plasmids in the dividing cell, the probability that none will enter this daughter is $(0.5)^n$. Since each division produces two daughter cells, the probability (P_0) that either one of them will not receive any plasmids (i.e. the segregation frequency) is given by:

$$P_0 = 2(0.5)^n = 2^{(1-n)}$$

A graph of P_0 against n is shown in Fig. 4.2. If there are more than 20 plasmids in a dividing cell, the segregation rate of 10^{-6} or less will

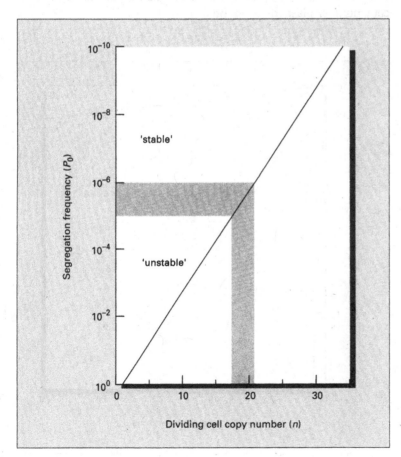

Fig. 4.2 The relationship between copy number and segregation frequency for randomly distributed plasmids. The shaded area represents the transition between 'stable' and 'unstable' plasmids based upon the sensitivity of conventional assays for plasmid loss.

be virtually undetectable. We might therefore expect all high copy number plasmids to be stably maintained without active partition. Although this is true for naturally occurring plasmids like ColE1 it is certainly not the case for the majority of cloning vectors.

4.3 Instability of randomly distributed plasmids

4.3.1 The kinetics of plasmid loss

Consider a randomly distributed plasmid with segregation frequency P_0. After one generation, a proportion P_0 of the cells are plasmid-free. After the second generation P_0 of the remaining plasmid-containing cells loose their plasmids, and so on. This is equivalent to first-order reaction kinetics and a graph where the log of the number of plasmid-containing cells is plotted against time gives a straight line. Figure 4.3 shows the theoretical kinetics of loss for plasmids with copy numbers from 2 to 20. It is immediately apparent that while random distribution can account for the high level of stability of natural, high copy number plasmids, active partitioning of low copy number plasmids is essential.

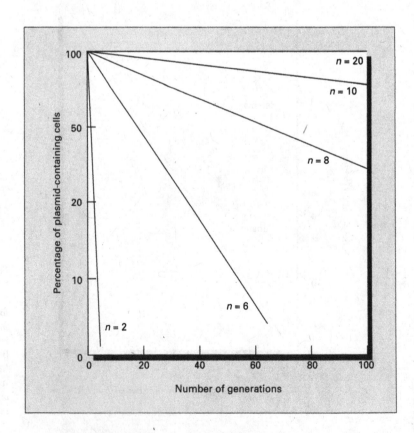

Fig. 4.3 Theoretical kinetics of loss for randomly distributed plasmids where *n* is the number of plasmids per dividing cell.

4.3.2 Metabolic load accelerates plasmid loss

In reality, the accumulation of plasmid-free cells is rarely consistent with first-order kinetics. Even if they arise infrequently, plasmid-free cells often accumulate rapidly because they outgrow their plasmid-bearing neighbours. In a study of 101 R factors from clinical isolates, Zund and Lebek (1980) found that one-quarter of them increased their host's generation time by more than 15%. There was some correlation between plasmid size and growth retardation which may reflect the effect of increased DNA replication load imposed by the plasmids. The sequestration of cell components for transcription and translation of plasmid genes can also have an adverse effect on cell growth. Even the toxicity of gene products may be significant when foreign genes are expressed in *Escherichia coli* or when chromosomal genes are over-expressed from a multicopy plasmid.

Figure 4.4 illustrates the kinetics of plasmid loss in a culture where plasmid-free segregants out-grow plasmid-bearing cells. The theoretical basis for this treatment was described by Boe *et al.* (1987). A characteristic feature of the system is an initial slow accumulation of plasmid-free cells followed by a rapid takeover of the culture. This pattern is seen when a culture of cells containing the cloning vector

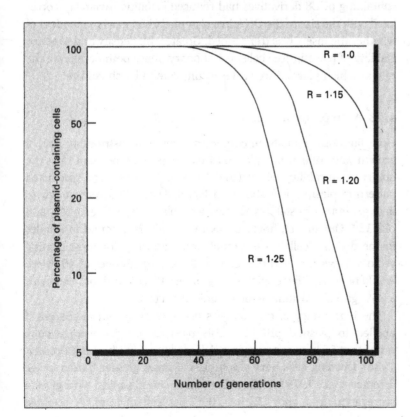

Fig. 4.4 Theoretical kinetics of loss for randomly distributed plasmids when plasmid-free cells out-grow plasmid-bearing cells. Dividing cells are assumed to contain 20 plasmids. R is the ratio of generation times for plasmid-bearing to plasmid-free cells.

pBR322 is grown in a chemostat under conditions of nutrient limitation (Jones *et al.*, 1980). The sudden disappearance of plasmid-bearing cells from chemostat cultures is known as washout and is a serious problem to biotechnologists because plasmid loss means a reduction in product yield.

Serial batch culture is a convenient alternative to the chemostat. Cells are grown to stationary phase, diluted into fresh medium and grown to stationary phase again. The cycle can be repeated as often as desired. In a chemostat the cells remain in a defined growth phase with a constant generation time but these parameters change constantly in batch culture. Consequently plasmid instability may vary with growth phase and be influenced as strongly by events in stationary phase as in exponential growth. Chiang and Bremer (1988) have analysed the stability of a series of pBR322-derived plasmids in serial batch culture. They found that although plasmid-free and plasmid-bearing strains had identical growth rates in exponential phase, the plasmid-free strain reached a higher cell density in stationary phase. Viability in stationary phase could also be reduced by the presence of a plasmid, particularly if it carried a tetracycline resistance gene. Similar observations were made by Chea *et al.* (1987) who found that stationary phase bacteria containing pUC8 derivatives had reduced viability. Inviability correlated with the size of the DNA insert in pUC8 but was independent of whether the insert was transcribed. Greater cell density and superior viability of plasmid-free cells in stationary phase both accelerate the rate at which plasmid-free cells accumulate in batch culture.

4.3.3 A simple numerical model of instability

How far does a model incorporating random distribution and a growth advantage for plasmid-free segregants account for the observed instability of plasmid cloning vectors? The predicted pattern of plasmid loss shown in Fig. 4.4 is qualitatively similar to data generated in stability studies with plasmids such as pUC8 and pBR322. Our model, however, assumes only 20 plasmid molecules in the dividing cell; a substantial under-estimate for most cloning vectors. If we repeated the analysis for a copy number of 40, there would be no significant plasmid loss after 100 generations, even with a 25% growth advantage for plasmid-free cells.

The inadequacy of our analysis becomes even more apparent if applied to plasmid pBR322. This plasmid has an average copy number of 55 in cells grown in a rich medium (Lin Chao and Bremer, 1986). Dividing cells have a volume 1.4 times greater than average (Bremer *et al.*, 1979) so should contain ≈ 80 plasmids. This gives a segregation frequency (P_0) of 10^{-24} per cell division. A scientist devoting his or her entire working life (about one million generations for *E. coli* in rich medium) to a plasmid stability experiment would still not be completely confident of detecting a plasmid-free segregant! In reality, derivatives of pBR322 are relatively unstable (Chiang and Bremer, 1988) indicating that plasmid-free cells arise at a much higher rate than the model predicts. Additional factors which promote instability must be missing from our model.

4.3.4 Copy number variance affects instability

There is little doubt that increased viability in stationary phase and faster growth in exponential phase are important influences on the rate at which plasmid-free cells accumulate. Nevertheless, these factors are not fundamental; they amplify instability but do not determine the rate at which plasmid-free cells arise in the first place. It is this latter question which merits the closest attention.

So far we have established that the rate at which plasmid-free cells arise is determined by the number of plasmids in the dividing cell. Our naïve analysis of pBR322 stability (section 3.3) uses a figure of 80 plasmids per cell. Experimentally determined copy numbers may be misleading, however, because they are mean values over a very large number of cells. If the copy number of individual dividing cells is variable, this may influence the rate of plasmid loss. Fig. 4.5 shows high and low variance copy number distributions for plasmids with a mean copy number 80. The plasmid with the low variance distribution will be extremely stable, but the plasmid with the high variance distribution will be less stable because plasmid-free cells arise from the low copy number end of the distribution at relatively high frequency. Instability will be detectable if a significant proportion of dividing cells have a copy number much less than 20, especially if plasmid-free segregants outgrow plasmid-bearing cells. Nevertheless, it would require a significant spread of copy number to explain the instability of pBR322 and, since copy number control systems exist to reduce variation, we need to consider whether this is a plausible cause of instability.

4.3.5 Factors affecting copy number variance

The copy number variance of dividing cells will depend, among other things, upon the efficiency with which uneven distribution of plasmids between daughters is corrected before the next cell division. Although the molecular basis of copy number control is well-known for an increasing number of plasmids, it is still unclear in most cases how tightly the number of replication events per generation is determined by the copy number of the new-born cell (i.e. the

71

physiology as opposed to the molecular biology of replication control). Without this knowledge, it is impossible to calculate the copy number variance in a population. Direct measurement is not possible because there is no reliable way to measure single-cell copy number.

Probably the most complete picture of the physiology of replication control exists for the actively partitioned, low copy number plasmid R1 (Nordström and Aagaard-Hansen, 1984; Nordström *et al.*, 1984). R1 replication is best described by a ' + *n*' model. This states that if *n* is the average copy number of new-born cells, an average of *n* plasmids will replicate in all cells during the subsequent generation, irrespective of the individual copy number. This pattern of replication is an inevitable consequence of the plasmid replication rate being inversely proportional to copy number. The actual number of plasmids replicated in an individual cell varies about the mean as a Poisson distribution. Assuming equipartition, the copy number variance among mother and daughter cells may then be calculated (Fig. 4.6). It turns out that the variance of copy number in dividing cells is not insignificant; 5% of dividing cells have a copy number less than half the average mean value.

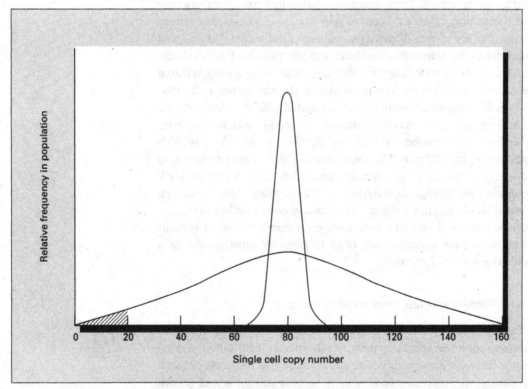

Fig. 4.5 Copy number distributions of plasmids with the same mean copy number but different variances. The hatched area represents a sub-population of low copy number cells from which plasmid-free segregants arise at high frequency.

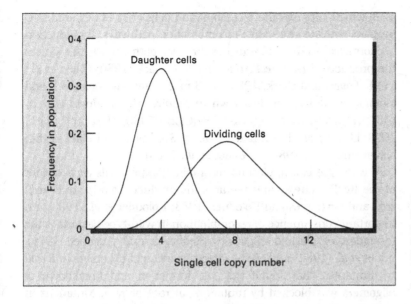

Fig. 4.6 Theoretical copy number distributions for plasmid R1 in dividing cells, and daughter cells immediately after division. From Nordström and Aagaard Hansen (1984).

We must be cautious when extrapolating this conclusion to other plasmids because the $+n$ model, with its constant replication rate per cell, is not universally applicable. In response to a two-fold increase in copy number the R1 replication rate is halved but plasmid F switches off replication altogether (Tsutsui and Matsubara, 1981). Studies with *Staphylococcus aureus* plasmid pT181 revealed unexpected behaviour at both high and low copy number. When pT181 copy number is recovering from an abnormally low level, the plasmid replicates faster than predicted by the model (Highlander and Novick, 1987) but when the RepC initiator protein is present in excess, replication is inhibited (Iordanescu, 1995).

There is no doubt that copy number variation is potentially of great importance in influencing the rate of plasmid loss. Unfortunately there is probably little to be gained by speculation about the extent and consequences of copy number variance until the physiology of replication control is better understood.

4.3.6 Host recombination systems affect plasmid stability

Plasmid oligomer formation is a major cause of instability through the generation of cells with abnormally low copy number. This section surveys the major *E. coli* recombination pathways and their activity on plasmid DNA. The seminal studies of Clark and co-workers led to the discovery of the RecBCD pathway of homologous recombination in *E. coli* (for reviews see Clark, 1973; Smith, 1988). This pathway is responsible for post-conjugational recombination and is inactivated by mutations in *recA*, *recB*, *recC* and *recD*. Two mutations (*sbcA* and *sbcB*) which suppress the recombination deficiency of *recB* and *recC* mutants activate alternative recombination pathways. By induction of exonuclease VIII, *sbcA* activates the RecE pathway which requires the products of *recA*, *recE recF*, and *recJ* (Clark, 1980; Gillen *et al.*, 1981; Lovett and Clark, 1984). *sbcB* inactivates exonuclease I, and stimulates the RecF pathway which requires the products of *recA*, *recF*, *recJ*, *recN*, *recO*, *recQ* and *ruv* (Clark, 1980; Horii and Clark, 1973; Lloyd *et al.*, 1983; Lloyd *et al.*, 1984; Lovett and Clark, 1984; Nakayama *et al.*, 1984; Kolodner *et al.*, 1985).

In wild-type *E. coli*, plasmid recombination proceeds *via* a variant of the RecF pathway and requires the products of *recA*, *recF*, *recJ*, *recO* and *ssb* (James and Kolodner, 1983; Kolodner *et al.*, 1985). An important consequence of recombination in wild-type bacteria is the formation of plasmid oligomers (Bedbrook and Ausubel, 1976). James *et al.* (1982) studied the interconversion of oligomers in *E. coli rec* mutants. They found that the formation and breakdown of oligomers was blocked by mutations in *recA* or *recF*. Mutations in *recB* or *recC* appeared to block oligomer formation (intermolecular recombination) but not breakdown (intramolecular events). The results are consistent with a model in which intermolecular recombination proceeds through a Holliday-type recombination intermediate. Processing of this intermediate results either in dimer formation or a return to monomers. Formation of a dimer requires the action of exonuclease V (the RecBCD gene product). In its absence, the intermediates are invariably resolved to monomers.

The proportion of plasmid oligomers in a culture is determined by two key factors; the intermolecular recombination rate and the growth rate of multimer-containing cells (Summers *et al.*, 1993). Surprisingly, the intramolecular recombination rate is of little significance. Once formed by recombination, a dimer replicates at twice the rate of plasmid monomers, leading to a clonal proliferation of higher forms known as a dimer catastrophe. Runaway multimerization is avoided because, for reasons which remain obscure, dimers reduce the host growth rate and the system eventually reaches equilibrium.

The equilibrium proportion of dimers is plasmid-specific and highly variable. For example, the cryptic plasmid p15A forms less than 10% oligomers while pACYC184 (in which the replication origin of p15A is joined to the tetracycline resistance gene of pSC101 and the chloramphenicol resistance gene of pKT002; Chang and Cohen, 1978) can be up to 95% oligomeric (James and Kolodner, 1983; James *et al.*, 1983). pACYC184 and other plasmids which readily form oligomers are thought to contain one or more DNA sequences which stimulate recombination by the RecF pathway in a manner analogous to *chi* in the RecBCD pathway (Stahl, 1979). In a search for recombino-

bacterial chromosome into pVH15 and screened for an increase in the proportion of oligomers. About 20% of inserts were active, increasing oligomer formation between eight- and 90-fold. Not surprisingly, oligomerization of cloning vectors containing large inserts is a common problem. Berg *et al.* (1989) reported a correlation between insert size and oligomer formation in pBR322 derivatives although it is unclear whether plasmid size itself is significant or whether larger inserts are simply more likely to contain recombinogenic sequences.

In addition to the RecF pathway, oligomers can also form by a *recA*-independent process. In *sbcA* mutants, induction of the RecE pathway results in a 20-fold stimulation of plasmid recombination and the formation of high levels of plasmid oligomers (Fishel, *et al.*, 1981; Laban and Cohen, 1981; James *et al.*, 1982; Cohen and Laban, 1983). Oligomer formation under these circumstances is sequence-independent. Intramolecular recombination in *sbcA* strains is largely unaffected by mutations in *recA* or *recF*, whereas intermolecular recombination is reduced. The effect on plasmid multimerization in an *sbcA* host can be dramatic; plasmid p15A has 7% oligomers in wild-type cells but this rises to 50% in an *sbcA* background (Fishel *et al.*, 1981). For this reason *sbcA* strains have often been used to investigate the link between oligomers plasmid instability.

4.3.7 Oligomers reduce plasmid stability

Mutations which affect plasmid recombination also influence stability. High copy number vectors such as pUC8, pBR322 and pACYC184 are most stable as monomers in recombination-deficient backgrounds, less stable in *rec*[+] and least stable in *sbcA*. The decrease in stability correlates with an increasing proportion of plasmid oligomers (Fig. 4.7; Summers and Sherratt, 1984). Two lines of evidence argue that changes in plasmid stability are due to alterations in levels of oligomers rather than a direct effect of the *rec* mutations. First, the stability of plasmids in *sbcA* mutants can be restored by inserting the Tn*3 res* site into the plasmid and supplying TnpR resolvase in *trans*. Resolvase mediates recombination between *res* sites in oligomers, converting them to monomers. Secondly, plasmid dimers 'frozen' in a *recF* background are extremely unstable compared to monomers in same background (Summers and Sherratt, 1984).

4.3.8 Plasmid oligomers have a reduced copy number

Why are plasmid dimers less stable than monomers? If dimers are maintained at a lower copy number and are distributed randomly at cell division, they will have a higher segregation frequency. In this context, copy number refers to the number of independent plasmid molecules, so a monomer or a dimer counts equally as one plasmid. Evidence that oligomers do, indeed, have a reduced copy number comes from direct comparison between monomer- and dimer-containing strains. Chiang and Bremer (1988) found that dimers of three different pBR322 derivatives were maintained at 45%, 50% and 67% of the monomer copy number. The copy number decrease is due to the increase in the number of origins rather than the increased size of dimers; a series of pUC8 derivatives containing tandem repeats of the origin region show a progressive drop in copy number as the number of origins increases (Summers and Sherratt, 1984). The evidence suggests that the replication control systems of ColE1 and related plasmids count origins rather than independent

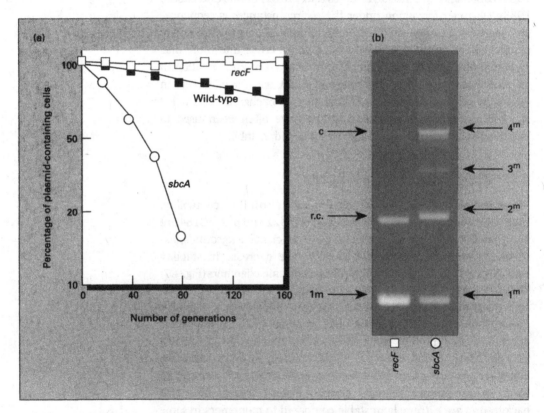

Fig. 4.7 The effect of oligomers on the stability of plasmid pBR322. (a) The plasmid is most stable in a *recF* strain (□), less stable in a recombination-proficient host (■) and very unstable in an *sbcA* strain (○). (b) Agarose gel electrophoresis reveals an absence of oligomers in the *recF* host. In addition to supercoiled monomers (1^m), slower-migrating bands corresponding to relaxed circular (r.c.) monomers and chromosome contamination (c) are visible. Extensive oligomerization is seen in the *sbcA* mutant. Bands corresponding to supercoiled plasmid monomers (1 ^m), dimers (2 ^m), trimers (3 ^m) and tetramers (4 ^m) are indicated.

plasmid molecules. A dimer is thus equivalent to two monomers and, in general, an oligomer of order n will have a copy number $1/n$ times the copy number of a monomer.

The effect of oligomerization on plasmid stability can be illustrated by comparing the theoretical segregation frequencies for monomer- and dimer-containing strains. Using the formula derived in section 4.2.2, a plasmid with a monomer copy number of 40 in dividing cells, has a segregation frequency of 2×10^{-12} ($= 2^{-39}$). If the dimer copy number is 20, the segregation frequency becomes 2×10^{-6} ($= 2^{-19}$); an increase of a million-fold simply as a result of dimerization.

4.3.9 Distribution of oligomers in a cell population

A dramatic difference in stability is seen between plasmids which are exclusively dimeric or monomeric. In wild-type cells, however, only 5–10% of plasmids are dimers. Is this sufficient to have a significant effect upon stability? At first sight, it seems unlikely. The segregation frequency for a plasmid with $n = 40$ is 1.8×10^{-12} ($= 2^{-39}$). Suppose that 10% of plasmids form dimers in a rec^+ background. On average each cell in the population will contain 36 monomers and two dimers; a total of 38 independent plasmid molecules. The segregation frequency now becomes 7.27×10^{-12} ($= 2^{-37}$). This increase of four-fold is not consistent with the data in Fig. 4.7 where a large difference in stability between a $recF$ and rec^+ host is associated with a small increase in the proportion of dimers.

The solution to this paradox may be to change our assumptions about the distribution of dimers in the cell population. Suppose that, instead of dimers being distributed evenly throughout the population, 10% of cells contain only dimers and 90% contain only monomers. The overall segregation frequency will be the weighted sum of the frequencies for monomer- and dimer-containing cells:

$$P_0 = 0.1 \times 2^{(1-20)} + 0.9 \times 2^{(1-40)}$$
$$\approx 0.1 \times 2^{(1-20)}$$
$$= 2 \times 10^{-7}$$

This is an increase of 10^5-fold over the monomer-containing strain and is far more consistent with the difference in stability of plasmids in rec^- and rec^+ strains.

Although this proposal provides a plausible explanation for the instability of many plasmids in rec^+ hosts it does not explain how this curious plasmid apartheid might arise. Consider a cell containing 40 monomers in which a single dimer arises as a result of a rare homologous recombination event. It has been known for some time that multicopy plasmid molecules are selected at random for replication (Bazaral and Helinski, 1970) but it is also true that origins are selected at random (Summers *et al.*, 1993). Thus in our cell, the probability of any one of the monomers being the next plasmid to replicate is 1/40, while the dimer has a replication probability of 2/40, simply because it contains two origins. Consequently dimers out-replicate monomers and, in the absence of counter-selection, there will be a rapid increase in the proportion of dimers. The outcome of this 'dimer catastrophe' is that approximately half of the dimers in a population are in dimer-only cells and the rate of plasmid loss is much greater than if the dimers were distributed evenly (Summers *et al.*, 1993).

4.3.10 Gaps in our understanding

In the absence of active partitioning, the crucial determinant of plasmid stability is the probability that a dividing cell will produce a plasmid-free daughter. Our problems begin when we attempt to estimate this probability for populations of cells because, although we can measure the mean copy number in a population, we have no clear idea of how it varies among individuals. Copy number variation will depend upon the physiology of the copy number control system and the formation and distribution of plasmid oligomers. These will influence the rate at which plasmid-free cells arise and additional factors such as differential growth rates and viability between plasmid-bearing and plasmid-free cells will have a major influence upon rate at which plasmids disappear from the culture. Although we have made significant progress in understanding the main causes of instability, the picture is by no means complete. With a mean copy number of ~ 200 in dividing cells, pUC8 should virtually never be lost, yet it is notoriously unstable. Perhaps the copy number control system operates poorly at such high copy number or plasmids cluster, thereby reducing the effective copy number. It seems that we will not have a complete picture of the causes of plasmid instability until we are able to determine the copy number and distribution of plasmids in individual cells.

4.4 Stability functions of natural high copy number plasmids

In contrast to synthetic cloning vectors, natural high copy number plasmids such as ColE1, pMB1 and p15A are extremely stable. In the absence of reliable evidence to the contrary, they are assumed to be distributed randomly at cell division. The majority of cloning vectors are derived from ColE1-related plasmids and their instability implies that functions which maximize stability have been lost during vector construction.

4.4.1 Sophisticated control circuits minimize copy number variance

Plasmid pMB1 has a copy number control system similar to ColE1 (see chapter 3). The primary repressor is a short transcript (RNA I) which hybridizes with the pre-primer RNA, alters its secondary structure and prevents its processing into a functional primer. A second level of control is provided by the Rom protein (Cesareni *et al.*, 1982) which stabilizes the binding of RNA I to its target, thus increasing its efficiency as a repressor. The pMB1 control circuit is intact in pBR322 (Bolivar *et al.*, 1977) but in its high copy number derivatives pAT153 and pUC8 (Twigg and Sherratt, 1980; Vieira and Messing, 1982), the *rom* gene has been deleted. Paradoxically, despite their increased copy number, these plasmids are less stable than pBR322 (D.K. Summers and D.J. Sherratt, unpublished data). Although the control systems of pBR322 and pUC8 maintain a constant mean copy number in the absence of the Rom protein, the decrease in stability suggests that the Rom may reduce the copy number variance by decreasing the response time of the control system to variations in copy number. Thus despite their high mean copy number, pAT153 and pUC8 could have an increased copy number variance with a 'tail' of low copy number cells from which plasmid-free segregants arise at high frequency. Although direct testing of this model is difficult, it seems inevitable that natural selection would favour control systems which minimize the copy number variance of randomly distributed plasmids.

4.4.2 Oligomers of natural plasmids are removed by site-specific recombination

The observation that ColE1 is stable and oligomers are rare, even in a hyper recombinogenic (*sbcA*) background, led to the discovery of a 240 bp plasmid site (*cer*) whose deletion simultaneously promotes oligomer formation and instability (Fig. 4.8). The *cer* site is a substrate for intramolecular site-specific recombination which converts oligomers to monomers (Summers and Sherratt, 1984; Fig. 4.9). Oligomer resolution sites are widespread among high copy number plasmids and have been identified in many plasmids including CloDF13 (Hakkaart *et al.*, 1984), pMB1 (Green *et al.*, 1981; Summers and Sherratt, 1985), ColK (Summers *et al.*, 1985), ColN (Kolot, 1990), ColA (Morlon *et al.*, 1988) and NTP16 (Cannon and Strike, 1992).

An unusual feature of *cer* and related oligomer resolution systems is that the proteins which mediate recombination are host-encoded. In most site-specific recombination systems, the recombinase is encoded by a gene adjacent to the recombination site. Mutations in four chromosomal genes abolish ColE1 oligomer resolution. These are *argR* (which encodes the repressor of arginine biosynthesis genes; Stirling *et al.*, 1988), *pepA* (the structural gene for aminopeptidase A; Stirling *et al.*, 1989) *xerC* (Colloms *et al.*, 1990) and *xerD* (Blakely *et al.*, 1993). XerC and XerD form a heterodimeric recombinase which binds to *cer* and mediates strand breakage and reunion during recombination. The same recombinases act at the *dif* site to resolve chromosome dimers as a prelude to partition (Blakely *et al.*, 1991).

cer-mediated recombination is severely constrained by the topological relationship of the recombining sites; dimer formation by intermolecular recombination is excluded, as is recombination between sites in inverted repeat. The imposition of topological constraint appears to be dependent upon the formation of a protein–DNA complex involving a full *cer* site, the recombinase and the two accessory proteins. A *cer* derivative which requires only the

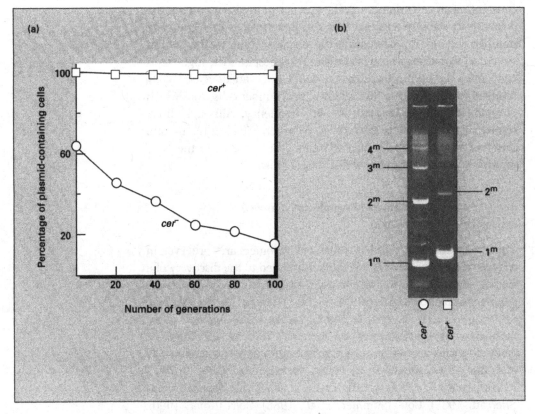

Fig. 4.8 Oligomer resolution and plasmid stability. (a) Stability of ColE1 derivatives, with and without a *cer* site, in an *sbcA* strain which produces plasmid oligomers. The *cer⁻* derivative is so unstable that nearly 40% of cells lost the plasmid during growth of the culture which was used to start the experiment. (b) Electrophoresis of plasmid DNA reveals oligomers of the *cer⁻* plasmid which are not seen for the plasmid containing the resolution site.

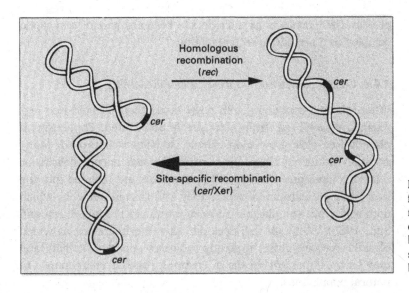

Homologous
recombination
(*rec*)

cer

cer

cer

cer

Site-specific recombination
(*cer*/Xer)

cer

Fig. 4.9 ColE1 oligomers formed by homologous recombination are converted to monomers by intermolecular, site-specific recombination between *cer* sites.

XerCD recombinase and a 50 bp fragment of the site to which the recombinase binds is no longer subject topological constraints and is equally capable of the formation and resolution of oligomers (Summers, 1989).

Oligomer formation by homologous recombination, proliferation by over-replication and removal by the slower growth of oligomer-containing cells (Summers *et al.*, 1993) establishes a dynamic equilibrium which *cer* displaces towards monomers. It is notable that *cer* stabilizes plasmids in an *sbcA* background even though significant numbers of plasmid oligomers are still present (Summers and Sherratt, 1984). For reasons explained in section 4.3.9, it is not necessary for *cer*-Xer recombination to remove oligomers altogether. It is sufficient to ensure that they are distributed evenly rather than being concentrated in a small proportion of cells. Oligomer resolution sites are nature's answer to the dimer catastrophe!

4.4.3 Regulation of cell division

Inactivation of a promoter buried in the middle of the ColE1 *cer* site reduces plasmid stability but has no effect upon oligomer resolution (Patient and Summers, 1993). The promoter directs the synthesis of a short transcript, Rcd, which inhibits cell division when over-expressed. There is evidence that transcription is elevated in oligomer-containing cells and it has been proposed that increased levels of Rcd prevent the division of these cells, which are at risk of producing a plasmid-free daughter. The mechanism by which the promoter senses whether it is in a monomer or an oligomer is, as yet, unclear but it is likely that it is activated by changes in the protein–DNA contacts as a single site becomes part of a synaptic complex as a prelude to recombination.

4.4.4 Colicin production improves plasmid stability

When plasmid-containing cells grow faster than plasmid-free segregants, plasmid loss from a culture is accelerated. Conversely, if plasmid-free cells grow more slowly, stability is improved. Many naturally occurring high copy number plasmids encode colicins or related compounds. These small polypeptides are released into the surrounding environment where they kill plasmid-free cells which do not produce the plasmid-encoded immunity product (Luria and Suit, 1987). Although this does not alter the frequency at which plasmid-free cells arise, it greatly increases apparent stability and may be an important means of ensuring plasmid maintenance in natural populations.

4.4.5 An integrated set of functions underpins multicopy plasmid stability

The formation of even a single dimer poses a major threat to multicopy plasmid stability. If it is not removed, the faster replication of the dimer means that dimer-only cells will arise within a few generations with the attendant high risk of plasmid loss. ColE1 musters a multifunctional response to the dimer threat. As dimers accumulate, a promoter within *cer* is activated and production of the Rcd transcript blocks cell division. Subsequently the dimers are resolved to monomers by site-specific recombination and the promoter returns to an inactive state. As the Rcd level in the cell falls, normal growth and division are restored. On the rare occasions that the system fails and a plasmid-free daughter arises, it will be killed by colicin produced by its siblings.

4.5 Active partition systems

4.5.1 Mechanisms and nomenclature

The stability of low copy number plasmids implies that they are partitioned actively at cell division (Jacob *et al.*, 1963) and it is generally assumed that this involves an association between plasmids and the cell membrane. The best-characterized partition functions are those of pSC101, the F factor, the P1 prophage (which behaves in all respects like a plasmid) and the IncFII drug resistance plasmids R1 and NR1 (R100). Our understanding of active partition has been reviewed by Austin (1988) and by Williams and Thomas

(1992). In principle, plasmids could simply hitch a ride with the host chromosome but the observation that mini-F (Ezaki *et al.*, 1991) and P1 (Funnell and Gagnier, 1995) are stable in hosts which are defective in chromosome partition argues against hitching as a universal mechanism.

Studies of plasmid replication control generally begin by mapping the minimal region required for autonomous replication. Plasmids consisting of a minimal replicon and a selectable marker are often highly unstable, being lost at a rate consistent with random distribution at cell division. Stable maintenance requires the presence in *cis* of an additional region (*par* or *stb*) encoding an active partitioning system. Both P1 and F have *par* regions immediately adjacent to the minimal replicon whereas the corresponding regions of R1 and NR1 are distant from the origin. Replication and partition are independent and the observation that *par* regions can stabilize unrelated plasmids (Ogura and Hiraga, 1983a; Gerdes and Molin, 1986; Austin *et al.*, 1986) implies that they must contain all necessary plasmid-encoded information.

The genetic nomenclature applied to partition systems can be confusing. In the past, *par* has been assigned to any locus at which mutation adversely affects plasmid maintenance. Thus the designation *parB* was applied both to the CloDF13 oligomer resolution function (subsequently renamed *crl*) and to the region of R1 which directs the killing of plasmid-free segregants. To avoid confusion in the future, it is important that *par* be restricted to loci which directly affect the distribution of plasmids into daughter cells.

4.5.2 The pSC101 partition locus

The genetic organization of the 375 bp *par* region of pSC101 (Meacock and Cohen, 1980) is quite different from its counterparts in F and P1 which are described in section 4.5.3. It does not appear to encode any protein but contains a binding site for DNA gyrase (Wahle and Kornberg, 1988). Some plasmids with partial deletions of *par* have decreased negative supercoiling and are extremely unstable. *topA* Mutations restore both negative supercoiling and stability to these plasmids. This has led to the proposal that gyrase-generated negative supercoiling establishes a DNA conformation which enables plasmids to interact with host structures which distribute them to daughter cells at division (Miller *et al.*, 1990). This may be an unnecessarily complicated interpretation of the data, however, because *topA* stabilization of partition-defective derivatives of F and P1 (Austin and Eichorn, 1992) appears to operate simply by restoring random plasmid distribution at cell division, possibly by facilitating separation of the products of replication.

An alternative interpretation of pSC101 *parB* function is suggested by parallels with the *cmp* locus of *S. aureus* plasmid pT181 (Gennaro and Novick, 1988), which stimulates binding of the RepC initiator protein to the origin. A role for *parB* in replication control is suggested by the observation that deletions in the locus affect plasmid copy number (Manen *et al.*, 1990). If loss of *parB* function reduces the efficiency with which copy number deviations are corrected, instability may be a consequence of an increased copy number variance rather than the loss of active partitioning (Nordström and Austin, 1989). Consistent with this idea, Conley and Cohen (1995) have recently shown that the biological effects of *par* correlate specifically with its ability to generate supercoils *in vivo* near the pSC101 replication origin.

4.5.3 Genetic organization of partition regions

The majority of partition loci are much larger and more complicated then pSC101 *parB*. The best-characterized of these belong to plasmid F and the P1 prophage Their genetic organization is similar; each encodes two *trans*-acting proteins and a *cis*-acting site (Mori *et al.*, 1986; Fig. 4.10). The genes for the partition proteins are expressed as a single operon which is subject to transcriptional autoregulation by the concerted action of its two products (Friedman and Austin, 1988). Co-ordinated expression of the proteins is essential for efficient partitioning; over-production of either of the P1 proteins causes severe plasmid instability (Abeles *et al.*, 1985; Funnell, 1988a). Host functions may also influence partitioning. The disruption of mini-F partitioning by mutations in *gyrB* is due to changes in plasmid superhelicity and consequent overproduction of SopB protein (Ogura *et al.*, 1990). It is probably a requirement for specific host-encoded components which makes the F partition locus work inefficiently outside *E. coli*, but a partition locus from RK2 (a broad host range plasmid) has been found to operate efficiently in *Pseudomonas*, *Azobacter* and *Agrobacterium* species (Roberts *et al.*, 1990).

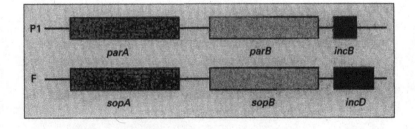

Fig. 4.10 Genetic organization of active partition cassettes of P1 and F.

4.5.4 The pre-pairing model

How do F and P1-like partition functions operate? A possible mechanism is described in the pre-pairing model of plasmid partitioning (Fig. 4.11; Austin and Abeles, 1983; Austin, 1988). A monomeric protein is proposed to bind a specific plasmid site. Binding causes the protein to dimerize, thus pairing the plasmids. The dimeric protein–DNA complex binds in its turn to a site in the membrane on the plane of cell division. Septum formation separates the plasmid pair and subsequent DNA replication releases the complex from the cell membrane.

The pre-pairing model is certainly plausible but can the products of the P1 partition locus be allocated roles in this process? The *cis*-acting site (*incB*) binds three molecules of the 39 kDa ParB protein. Two molecules bind to a perfect 13 bp palindrome and the third binds to a region separated from the palindrome by a recognition site for integration host factor (Davis and Austin, 1988; Funnell, 1988b). Integration host factor (IHF) is a host-encoded heterodimeric protein implicated in DNA bending in a range of condensed nucleoprotein complexes (reviewed by Friedman, 1988 and Freundlich *et al.*, 1992). If the pre-pairing model is correct, ParB must be the plasmid-pairing protein and *incB* its recognition site. An

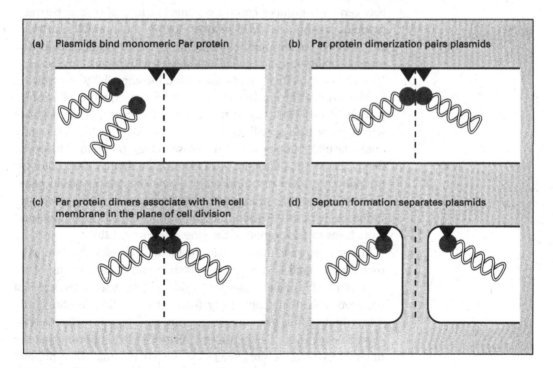

(a) Plasmids bind monomeric Par protein

(b) Par protein dimerization pairs plasmids

(c) Par protein dimers associate with the cell membrane in the plane of cell division

(d) Septum formation separates plasmids

Fig. 4.11 The pre-pairing model of plasmid partition.

observation that SopB (the analogue of ParB encoded by F) sediments with the cell membrane fragment under certain conditions (Watanabe *et al.*, 1989) implies that these proteins may also anchor paired plasmids to the membrane.

ParA binds to the *parA–parB* operator but not to *incB*, so is unlikely to be involved in plasmid pairing. It is a member of a family of ATPases which includes a membrane-associated protein involved in arsenate resistance (ArsA), an inhibitor of *E. coli* cell division (MinD) and a variety of partition proteins including RK2 IncC and F SopA (Motallebi-Veshareh *et al.*, 1990; Williams and Thomas, 1992). The similarity between MinD and ParA hints at a fundamental similarity between plasmid and chromosome partitioning. Possibly ATP hydrolysis by the ParA–ParB–*incB* complex provides the energy to separate the products of plasmid replication in a process analogous to mitosis in eukaryotic cells. The ATPase activity of ParA is strongly stimulated by ParB and DNA (Davis *et al.*, 1992) and it may form part of a membrane recognition site for paired plasmids. Plasmid movement could result from the translocation of DNA past a ParA complex anchored in the cell membrane.

4.5.5 Partition-related incompatibility

Plasmid incompatibility is most often a consequence of interference between copy number control systems but may also arise between plasmids having only their partition loci in common (Nordström *et al.*, 1980b; Austin and Nordström, 1990). Mutations affecting incompatibility map to the *cis*-acting sites within partition loci (e.g. P1 *incB* and F *incD*). Partition-associated incompatibility is predicted by the pre-pairing model and is a direct consequence of plasmid pairing. If two plasmids in the same cell encode pairing proteins which show high specificity for their respective binding sites, only homoplasmid pairs form and the plasmids are partitioned independently (Fig. 4.12a). Plasmids with identical or closely related sites for the pairing protein can form either homo- or heteroplasmid pairs. Depending upon the orientation of the paired plasmids with respect to the plane of cell division, the partitioning of some heteroplasmid pairs leads to segregation of the plasmids (Fig. 4.12b).

Reassessment of the pre-pairing model followed the discovery that the minimal *cis*-acting site which supports partitioning of P1 or F is much smaller than the region required for expression of partition-related incompatibility (Lane *et al.*, 1987; Martin *et al.*, 1987). In its simple form, the pre-pairing model predicts that the two should be identical. Furthermore it seemed to be significant that the minimal P1 partition site (*parS*) lacks the IHF binding site. A plasmid carrying this site is compatible with a plasmid containing the full incompatibility determinant (*incB*) in a wild-type host, but incompatibility is restored in an IHF-deficient mutant (Davis and Austin, 1988; Davis *et al.*, 1990). A selective pairing mechanism was suggested to explain these observations. It is proposed that a *parS*–ParB complex is structurally distinct from an *incB*–IHF–ParB complex because of the bending of DNA by IHF. These different complexes cannot pair, so expression of incompatibility (which requires the formation of heteroplasmid pairs) is blocked. In the absence of IHF, the *incB*–ParB and *parS*–ParB complexes are sufficiently similar that heteroplasmid pair formation (and hence incompatibility) is restored.

Although partition-associated incompatibility follows logically from the pre-pairing model, it is not an inevitable consequence of active partitioning. A notable exception is plasmid NR1; two plasmids sharing only the NR1 partition locus (*stb*) are compatible.(Min *et al.*, 1988; Tabuchi *et al.*, 1988). This may mean that pairing of NR1 plasmids follows replication immediately and the plasmids are not selected from an unpaired pool as appears to be the case for P1 and F.

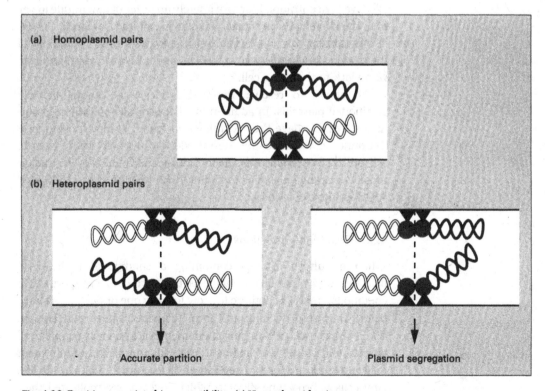

Fig. 4.12 Partition-associated incompatibility. (a) Homoplasmid pairs are partitioned accurately. (b) Formation of heteroplasmid pairs results either in accurate partition or plasmid segregation.

4.5.6 Dimer resolution assists active partitioning

The P1 prophage plasmid is lost at less than one in 10^4 cell divisions, implying that replication and partitioning are both highly efficient processes. In recombination-proficient hosts, however, some par^+ P1 mini-plasmids are lost at one in a hundred cell divisions (Austin *et al.*, 1981). This instability is largely elimenated in a rec^- background and is due to homologous recombination between P1 monomers following replication. The product of recombination is a dimer which cannot be partitioned. Unstable P1 mini-plasmids lack the *loxP-cre* region which encodes a site-specific recombination system. Cre recombinase recombines the directly repeated *lox* sites in a dimer restoring monomers which can be partitioned successfully.

In vitro, *lox*-Cre is promiscuous, and supports both the formation and breakdown of dimers. For this reason it has been proposed that the constant, rapid interconversion of monomers and dimers may be sufficient to provide monomers for partition, regardless of the whims of homologous recombination. *In vivo*, however, the system is more restrained and dimer breakdown predominates (Adams *et al.*, 1992). It appears that *in vivo* DNA accessibility is restricted and that the effective concentration is approximately an order of magnitude lower than the chemical concentration (Hildebrandt and Cozzarelli, 1995). The reason for this is not clear but it provides a timely reminder of the dangers inherent in the uncritical application of results from *in vitro* work to the living cell.

The ability to resolve dimers is widespread among actively partitioned plasmids. In addition to P1 *lox-cre*, site-specific recombination systems have been identified on many plasmids including F (O'Connor *et al.*, 1986) and R46 (Dodd and Bennett, 1986) from *E. coli*, pSDL2 from *Salmonella dublin* (Krause and Guiney, 1991) and on broad host range plasmids including RK2 (Roberts *et al.*, 1990) and RP4 (Gerlitz *et al.*, 1990).

4.5.7 Host killing functions

When a culture of cells containing a conditional replication–defective derivative of plasmid F is transferred to non-permissive conditions, the cells continue to grow for only one or two generations before they stop dividing and form filaments (Ogura and Hiraga, 1983b; Miki *et al.*, 1984). This phenomenon is known as coupled cell division (*ccd*) and it was assumed originally to reflect a requirement for completion of plasmid replication before an F$^+$ cell can divide. Any cells approaching division which had not yet replicated their plasmid, were thought to stop growing and allow the plasmid to catch up, thus avoiding the production of a plasmid-free daughter.

The implicit assumption of symbiosis between plasmid and host has since proved rather naïve.

The coupled cell division model was abandoned after it was found that cells containing a single copy of F are capable of division (Jaffe *et al.*, 1985; Hiraga *et al.*, 1986). The products of division are a viable plasmid-containing cell and a plasmid-free cell which, after a few residual divisions, forms filaments and dies. Ccd thus increases the *apparent* stability of F by post-segregational killing of plasmid-free segregants. The host-killing function consists of the products of *ccdA* and *ccdB* which form part of an autoregulated operon (de Feyter *et al.*, 1989; Tam and Kline, 1989) and encode polypeptides of 8.3 and 11.7 kDa, respectively (Bex *et al.*, 1983; Miki *et al.*, 1984). In plasmid-containing cells, CcdA serves as an antidote to the effect of CcdB, possibly by binding to it (Tam and Kline, 1989). In the absence of CcdA, CcdB prevents proper partitioning of the host cell chromosome by inhibiting the activity of DNA gyrase (Miki *et al.*, 1992) or by causing gyrase-induced double-stranded breaks in the host chromosome (Bernard and Couturier, 1992).

Newly formed plasmid-free cells initially contain both CcdA and CcdB. In the absence of further transcription of the *ccd* operon, the concentration of both polypeptides begins to fall. Differential stability of poison and antidote are the key to post-segregational killing. CcdA is unstable and is degraded by the Lon protease (Van Melderen *et al.*, 1994) allowing the more stable CcdB to exert its killing action. A related host-killing system has been described for plasmid R100 (also known as NR1; Tsuchimoto *et al.*, 1988).

A functionally analogous but mechanistically distinct host killing system is encoded by the *parB* locus of plasmid R1. Cells which lose R1 become inviable, appearing as transparent cell 'ghosts' under the phase contrast microscope (Gerdes *et al.*, 1986a). Genetic analysis initially identified two genes (*hok* and *sok*) within *parB*. They are encoded on opposite strands with a 128 bp overlap at their 5′ ends. The Hok polypeptide is responsible for host cell killing, while *sok* encodes a *trans*-acting RNA which protects plasmid-containing cells against Hok by blocking translation of *hok* mRNA. Hok is a membrane-associated polypeptide of 50 amino acids whose expression leads to a loss of cell membrane potential, arrest of respiration, changes in cell morphology and death (Gerdes *et al.*, 1986b). It has 40% homology with the product of *E. coli relF* (Bech *et al.*, 1985) and is related to a set of proteins, conserved among a wide range of bacterial species, whose over expression leads to cell death (Poulsen *et al.*, 1989). It looks as if R1 has hijacked a chromosomal gene for its own selfish purposes.

The mechanism by which Sok RNA blocks translation of Hok in plasmid-containing cells became apparent only after the identification of *mok*, a third gene in the *parB* locus whose open reading frame starts about 150 bp upstream of *hok* (Fig. 4.13; Thistead and Gerdes, 1992). *mok* and *hok* are co-transcribed, and translation of *hok* is completely dependent upon translation of *mok*. Sok antisense RNA forms a duplex with the 5′ end of the *mok–hok* message rendering the *mok* ribosome binding site inaccessible to ribosomes and promoting RNase III cleavage and degradation of the mRNA (Gerdes *et al.*, 1992). In the absence of *mok* translation, *hok* is not expressed from intact message, even though its own ribosome binding site is not directly obscured by Sok RNA. When a plasmid-free cell is formed, the unstable Sok RNA decays much more rapidly than the stable

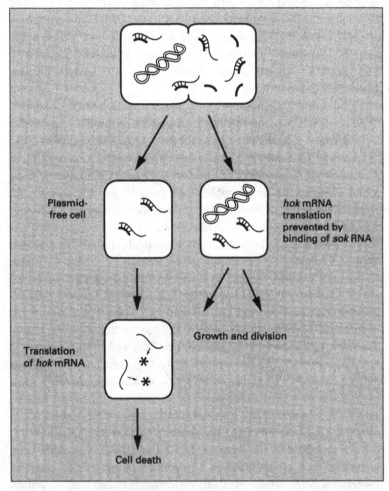

Fig. 4.13 Post-segregational killing by the *parB* (*hok–sok*) system of plasmid R1. A plasmid-containing cell contains *hok* mRNA but translation is blocked by Sok antisense RNA. In a plasmid-free cell, unstable Sok is degraded and translation of *hok* mRNA leads to cell death. In the interests of clarity the role of *mok* (the third gene in the *parB* cassette) has been omitted. From Gerdes *et al.* (1988).

mok–hok message ($t_{1/2} \approx 20$ min) and, as the protection afforded by Sok is lost, Mok and Hok are translated and the cell dies (Gerdes *et al.*, 1988; Thistead and Gerdes, 1992).

4.5.8 Belt and braces killing systems

Host killing systems are widespread among low copy plasmids. In addition to F *ccd* and R1 *parB* (*hok–sok*) a third class of host-killing (or 'plasmid addiction') system has been identified in bacteriophage P1. It contributes to the extreme stability with which the plasmid-like prophage is maintained (Lehnherr *et al.*, 1993). The lack of significant DNA homology among these systems and the differences in mechanism suggest that they are of independent evolutionary origin.

An unexpected postscript to this story is the discovery that many plasmids encode multiple host-killing functions. In addition to *ccd*, plasmid F contains a host-killing function (*stm* or *flm*) whose DNA sequence is almost identical to R1 *parB* (Golub and Panzer, 1988; Loh *et al.*, 1988). *parB*-like systems appear to be widespread and there is evidence for homologous regions in plasmids R100, R124, pGL611A, pIP231 and pIP71a (Golub and Panzer, 1988; Gerdes *et al.*, 1990). Plasmids R1 and R100 were found to carry, in addition to *parB*, related host-killing loci (*parD* and *pem* respectively) which promote plasmid stability through the antagonistic action of proteins analogous to CcdA and CcdB of plasmid F (Bravo *et al.*, 1987; Bravo *et al.*, 1988; Tsuchimoto and Ohtsubo, 1989; Tsuchimoto *et al.*, 1992). *parD* activity is repressed in the wild-type plasmid but it can be activated by mutation. Despite the similarities of genetic organization and limited amino acid sequence homologies between *parD* and *ccd* they do not appear to be closely related (Ruiz-Echevarria *et al.*, 1991).

What is the significance of multiple host killing systems? A clue lies in reports that individually they are not 100% effective; if only one is induced, some cells may live to tell the tale. Multiple systems decrease significantly the chance that any cells will survive plasmid loss.

4.5.9 Concerted action of stability functions

Rather like their high copy counterparts, the stable maintenance of low copy plasmids results from the concerted action of a series of distinct stability functions. Accurate control of replication is essential to ensure that at least two plasmid copies are available for active partition in a dividing cell. The threat of dimer formation is met by plasmid-encoded recombination systems which restore plasmids to the monomeric state. If despite all these preventative measures a plasmid-free cell is formed, host-killing functions ensure that the segregant has a very bleak future indeed.

5: Plasmid Dissemination

5.1 Mechanisms of plasmid spread

5.1.1 Plasmid ecology

Plasmids play a crucial role in bacterial evolution by providing a massive reservoir of genetic information which is accessible to virtually all species in all habitats. The widespread sharing of genes among bacteria is possible because plasmids are capable of cell-to-cell transfer across species and generic boundaries. In a first attempt to understand plasmid ecology, we will address the mechanisms and the efficiency of the transfer processes. Plasmids can be transferred from cell to cell in three ways: passively by transduction or transformation, and actively by conjugation. It has generally been assumed that conjugation is of greater evolutionary significance than transformation or transduction. Recently, however, concern over the possible escape and proliferation of engineered extrachromosomal elements has led to a re-evaluation of the importance of transformation and transduction in the natural environment.

5.2 Transduction

In the early 1950s, Lederberg and co-workers observed gene transfer between strains of *Salmonella typhimurium* when one strain was exposed to a cell-free extract of another (Lederberg *et al.*, 1951). This process, which they termed transduction, was mediated by bacteriophage particles which acted as vectors to carry DNA between strains (Zinder and Lederberg, 1952). Generalized and specialized transduction have since proved invaluable in microbial genetic research. Generalized transduction, as mediated by *E. coli* phage P1, provides a means to move any chromosomal gene betwen hosts, as long as appropriate selection is available. Gene transfer by specialized transduction is less generally useful because it is restricted to genes which lie adjacent to the integration site of a temperate phage such as λ. This section considers generalized transduction as a potential mechanism of plasmid gene transfer.

5.2.1 Generalized transduction

Generalized transduction has the potential to transfer both plasmids and chromosome fragments. Transducing particles are produced by accidental packaging of host DNA in a phage head; the size and shape of the head determines the amount of DNA transfered. The length of DNA packaged is strictly constrained; it must not be too large or too small for the head and these size limits are significant when considering the transduction of plasmid DNA. Saye *et al.* (1987) found that plasmid transduction by isolates of *Pseudomonas aeruginosa* phage F116 was more efficient for plasmids similar in size to the phage genome. Novick *et al.* (1986) found that small plasmids of *Staphylococcus aureus* could be transduced only because they form high molecular weight linear concatemers in the host cell.

As long as packaging conditions are satisfied, plasmid-borne genes have the potential to be transduced more efficiently than chromosomal determinants. If donor and recipient have significant genetic differences, there may be insufficient homology to support integration of transduced fragments into the chromosome. A window of opportunity for transduced DNA to integrate into the chromosome arises soon after injection and, once passed, does not recur. As a result, most chromosome fragments give rise to abortive transductants. The injected DNA is not integrated and, although it may persist for a while in the recipient cell, it is incapable of autonomous replication and is subject to degradation. In contrast, integration is not required for the survival of transduced plasmid DNA. As long as the new host provides the necessary replication functions, it is necessary only that the plasmid recircularizes to resume its autonomous existence.

5.2.2 Transduction in nature

Generalized transducing phages have been identified in over 20 different bacterial genera (Kokjohn and Miller, 1992) but bacteriophages often have a narrow host range and it may seem likely that transduction is a less important means of gene transfer than conjugation. Nevertheless, substantial concentrations of bacteriophages have been reported in natural environments including freshwater habitats (10^3–10^4 plaque-forming units (pfu) ml^{-1}), waste water, sludge and sewage (10^8–10^{10} pfu ml^{-1}), and marine waters (10^5–10^{11} pfu ml^{-1}). Transduction of plasmid DNA has been demonstrated in soil, fresh water and synthetic waste-water environments (Kokjohn and Miller, 1992). Furthermore, although phage infection often requires an interaction between phage proteins and specific receptors on the host surface, numerous examples of phage with host ranges which cross species and generic barriers have been reported. These include phages PRR1 and PRD1 which will infect any Gram-negative bacterium containing an IncP-1 plasmid (which encodes the phage-specific receptor) and others which mediate gene transfer among the staphylococci and the enterobacteriaceae. Clearly there exists a potential for gene exchange by transduction in natural environments and Novick *et al.* (1986) have gone so far as to suggest that it may be the major mechanism for genetic dissemination in some species.

5.3 Transformation

Transformation is the process whereby a cell takes-up and expresses exogenous DNA and was the first mechanism of bacterial gene exchange to be described (Griffith, 1928). Although this phenomenon is exhibited by a wide range of bacteria, it has been studied in depth for only a few model systems. When considering the role of plasmid transformation in bacterial evolution, the frequency of transformation which can be achieved by unnatural means such as electroporation, protoplast fusion and $CaCl_2$ treatment are irrelevant; we need to establish the potential for gene transfer by transformation in natural environments (Stewart, 1992).

5.3.1 The mechanism of transformation

Crucial stages in the process of transformation are the acquisition of competence (the potential to be transformed), binding of DNA to the cell surface, uptake and chromosomal integration or recircularization of the transforming molecule. In *Bacillus subtilis*, *Streptococcus pneumoniae* and certain other Gram-positive organisms, competence is induced by the accumulation in the local environment of a low molecular weight protein (a competence factor). There is no corresponding exogenous factor for Gram-negative species such as *Haemophilus influenzae*, *Pseudomonas stutzeri* and *Azotobacter vinelandii* where the induction of competence is associated with a change in growth rate or a period of unbalanced growth at the transition between exponential growth and stationary phase. During the development of competence, the cells express genes for membrane-associated DNA binding proteins, which show a strong preference for double-stranded DNA (as opposed to single-stranded DNA or RNA). While *Streptococcus* and *Bacillus* display no preference for self or non-self DNA, *Haemophilus* shows a specificity for DNA from the same, or closely related species. Tight binding to the cell surface is associated with the introduction of multiple single-stranded breaks in the DNA which enters the cell as a single strand. Hydrolysis of the excluded strand may provide energy to drive uptake. Subsequent integration of transforming DNA into the recipient chromosome requires the enzymes of generalized recombination although this step can be avoided by plasmid DNA which needs only to recircularize if it is capable of autonomous replication in the new host.

5.3.2 Transformation in nature

The likelihood of transformation in natural environments has been discussed by Stewart (1992). He concludes that although the potential for natural transformation exists in many environments, we are still unclear of its relative importance as a means of horizontal gene flow. Laboratory studies show that competence is usually repressed during exponential growth but, in natural environments, organisms are likely to be slow-growing or in stationary phase for extended periods. It is possible that natural competence is more common than we expect and the need for special conditions to induce competence may be a laboratory artefact. The composition of the microenvironment is sure to influence the efficiency of transformation by affecting the stability of extracellular DNA and the frequency and duration for which strains become competent. Key factors will include the physical and chemical composition of the growth medium, including the ionic concentration and pH. Binding of the DNA and recipient cells to surfaces where the nucleic acid is protected from degradative enzymes is likely to be important, as are conditions which restrict the loss by diffusion of exogenous competence factor.

Even. if the frequency of transformation is small, a cell whose fitness is increased by the uptake of DNA (e.g. through an increase in metabolic efficiency or the acquisition of resistance to environmental toxins) will out-grow its neighbours thereby increasing the frequency of the newly acquired gene. This, in turn, increases the chance that it will be passed on to other cells by conjugation, transduction or by transformation. Natural selection may also change the ability of the bacterial population to take up exogenous DNA. If variation in competence levels has a heritable component, the most competent cells will be most likely to benefit from the increase in fitness associated with the acquisition of a new, favourable genotype and will increase in frequency within the population.

5.4 Conjugation in Gram-negative organisms

During conjugation, DNA is transferred between bacteria by a mechanism requiring cell-to-cell contact (Lederberg and Tatum, 1946). The nucleic acid never leaves the protection of the cellular environment and is thus protected from degradation by extracellular nucleases and heavy metals. Recently interest in the mechanism of conjugation has been revived with the realization that it can promote gene exchange among an enormous variety of bacterial species and genera.

The information required for transfer is encoded on transfer-proficient (Tra$^+$) plasmids which may also be able to promote the transmission of mobilizable plasmids and even chromosomal genes. Numerous genetically distinct and non-interacting conjugation systems have been identified in Gram-negative and Gram-positive organisms. Conjugation is typically associated with an extra round of plasmid replication (Fig. 5.1) and so provides a mechanism for the plasmid to proliferate more rapidly than its host. In genetic terms the capacity for self-transmission is relatively costly and, to a large extent, constrains the life-style of the plasmid. For example plasmid F devotes 33 kb to transfer functions which means that it must maintain a low copy number to avoid imposing an excessive

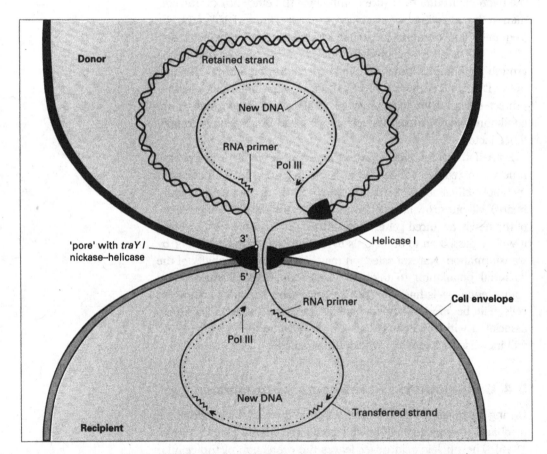

Fig. 5.1 Plasmid transfer by conjugation. A single strand of the plasmid is passed through a pore between the donor and recipient. DNA synthesis in both donor and recipient means that conjugation is accompanied by an increase in the total number of plasmid copies. From Willetts (1985).

metabolic load. Low copy number demands an active partition system which means a further increase in size. Overviews of bacterial conjugation have been produced by Ippen-Ihler (1989) and Day and Fry (1992). A more specific description of the F plasmid paradigm is provided by Willetts and Skurray (1987).

5.4.1 Mating pair formation

Conjugative plasmids from Gram-negative bacteria direct the synthesis of an extracellular pilus, which has an essential role in the recognition of recipient cells and the establishment of cell-to-cell contact. Pili can be classified into two broad morphological groups: long flexible (1 μm) and short rigid (0.1 μm). Long pili, like those expressed by cells carrying the F plasmid, support genetic exchange on surfaces and in liquid media at roughly equal frequencies. Short pili expressed by plasmids of the IncN, P and W incompatibility groups are 10^3–10^5 times more efficient in surface matings than in liquid. Some plasmids (Inc groups I_1, I_2, I_5, B, K and Z) encode both long and short pili. Current evidence indicates that pili are not involved in conjugation between Gram-positive organisms.

Of the many plasmid conjugation systems in Gram-negative bacteria, only those of the IncF plasmids have been studied in detail. Until relatively recently it was thought that the host range of plasmid F was limited to a few members of the Enterobacteriacae and the observation that it could promote gene transfer between *Escherichia coli* and yeast (Heinemann and Sprague, 1989) came as a surprise. General reviews of F conjugation include Willetts and Skurray (1987) and Ippen-Ihler and Minkley (1986) and references to more recent papers on specific topics are given below. The genes required for conjugation (more than 30 have been identified to date) are clustered in the 33 kb *tra* region which constitutes about one-third of the plasmid. Their products accomplish all transfer-related functions which include: (a) surface exclusion (a *traT* and *traS*-dependent process, which blocks transfer to a cell that possesses the plasmid already; Harrison *et al.*, 1992); (b) synthesis and assembly of the F-pilus (Maneewannakul *et al.*, 1992a); (c) stabilization of cell aggregates formed during mating; and (d) nicking unwinding and transport of the transferred strand to the recipient cell (Firth and Skurray, 1992; Maneewannakul *et al.*, 1992b).

At least 14 *tra* genes are involved in construction of the F-pilus; a hollow cylinder of 8 nm diameter with a 2 nm axial hole. The pilus is constructed from pilin subunits which are encoded by *traA* and processed by the product of *traQ*. Mutations in *traL, E, K, B, V, C, W, U, F, H* and *G* lead to the accumulation of pilin in the inner membrane, implying that these genes are involved in pilus assembly.

After the initial contact between the tip of the pilus and the recipient cell the pilus retracts, bringing the donor and recipient into close contact. Current models suggest that pilus disassembly is of crucial importance in the establishment of contact between the donor and recipient cell surfaces and the formation of a DNA transport pore.

5.4.2 Plasmid transfer

In the second stage of conjugation a single-strand nick is introduced at the origin of transfer and the nicked strand is transferred to the recipient (Fig. 5.1). The non-transferred strand is copied so the donor retains an intact plasmid. During transfer the nicking protein remains bound covalently to the exposed 5' terminus so the process is analogous in many ways to the replication of pT181-like plasmids in *Staph. aureus* (Fig. 3.8).

The specificity of transfer initiation lies in the recognition and nicking of *oriT*. A wide ranging discussion of the structure and function of transfer origins can be found in the review by Waters and Guiney (1993). In plasmid F the 250 bp *oriT* locus contains sites for the binding of TraY, TraI TraM and IHF. Nicking *in vivo* requires proteins TraY and TraI; the 5'-3' helicase activity of TraI is responsible for unwinding the donor duplex to facilitate transfer. TraM binds to *oriT* (Di Laurenzio *et al.*, 1992) and is thought to sense mating pair formation and trigger the initiation of conjugative DNA synthesis. A complement to the transferred strand is synthesized in the recipient, probably by rolling circle replication. This requires the chromosome-encoded enzymes of DNA replication including DNA polymerase III holoenzyme. Recircularization of the transferred plasmid is *oriT*-dependent but independent of recombination systems encoded by the recipient, so it is unlikely to involve recombination within multimeric linear plasmids. An attractive hypothesis for the mechanism of recircularization is that the nicking complex bound to the transferred 5' end is retained at the membrane pore where it interacts with the trailing 3' end and religation occurs by a reversal of the nicking process (Willetts and Skurray, 1987). Mating is followed by active disaggregation of donor and recipient cells.

5.4.3 Regulation of conjugative transfer

The majority of conjugal transfer systems are tightly regulated and plasmid transfer is rare. This may reflect the hazards of pilus expression which include sensitivity to a multiplicity of male-specific phages. Significantly, the arrival of a conjugal plasmid in a new host is followed by a period when its transfer system is transiently derepressed, increasing the chance that it will transfer again to a plasmid-free neighbour. This initiates an epidemic spread of the plasmid through the local environment and Lundquist and Levin (1986) have proposed that such transitory derepression may be sufficient to maintain conjugative plasmids even in the absence of direct selection.

The *tra* genes of F-like plasmids (including R100) are under positive regulation by the product of *traJ*. Translation of *traJ* mRNA is repressed by the concerted action of FinP and FinO (Fig. 5.2) which together reduce the frequency of conjugation 1000-fold. FinP is an 80 nucleotide antisense RNA, complementary to the first two codons and part of the 5′ untranslated region of the *traJ* message. Frost *et al.* (1989) suggested that FinO does not directly repress transcription of the *tra* operon but protects the FinP transcript from degradation and facilitates its interaction with *traJ* mRNA. This was confirmed by a demonstration that FinP is stabilized *in vivo* by the 22 kDa FinO protein (Lee *et al.*, 1992); its half life is increased from three minutes to over 40 minutes. More recently, van Biesen and Frost (1994) have reported that FinO binds to one of the two stem loops of FinP and to the complementary structure in the *traJ* mRNA. This interaction not only protects FinP RNA from degradation but also increases the rate of FinP–*traJ* mRNA duplex formation five-fold. Although the *tra* genes of most conjugative plasmids are under strict regulation,

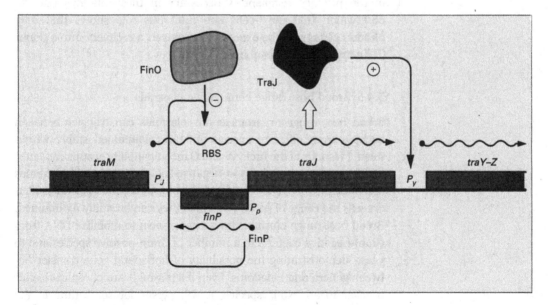

Fig. 5.2 Control of genes involved in conjugation. P_P, P_J and P_Y represent promoters for the *finP*, *traJ* and *traY* transcription units. Transcription of the *tra* operon is activated by TraJ. Regulation of *traJ* is post-transcriptional; antisense RNA FinP obscures the ribosome binding site (RBS) of *traJ* mRNA. In most plasmids FinP is assisted by the FinO protein but in plasmid F, *finO* is inactivated by a transposon insertion. After van Biesen *et al.* (1993).

plasmid F is naturally derepressed by virtue of an IS3 element inserted within *finO*. Full repression is seen only when the gene product is supplied in *trans* from a related plasmid. Despite the complexity of the control circuits described here, the story is not yet complete as mutations in several host genes including *sfrA*, *sfrB*, *cpxA* and *cpxB*. also affect the frequency of transfer.

5.4.4 Retrotransfer

Although conjugation is described commonly as bacterial sex, it is normally a parasexual process involving a unidirectional transfer of information; the genotype of the donor remains unchanged by the mating. Mergeay *et al.* (1987) reported a more equitable variation on conjugation, known as retrotransfer or shuttle transfer. Retrotransfer is mediated by broad host range plasmids of the IncP1, IncM and IncN groups. It is a sequential process in which the acquisition of a conjugative plasmid converts the recipient into a donor. The result is a reshuffling of plasmid and chromosomal genes between the mating pair. Retrotransfer of chromosomal markers mediated by IncP1 plasmid pULB113 has been seen in single-species matings of *P. fluorescens*, *Alcaligenes eutrophus* and *S. typhimurium*, and heterospecific matings between *A. eutrophus* and *P. fluorescens*. Gene expression in the primary recipient is necessary in these matings but the observation that maxi-cells can fulfil this role shows that only plasmid genes need be expressed to convert a recipient into a donor (Heinemann and Ankenbauer, 1993).

5.4.5 Broad host range conjugative plasmids

Broad host range or promiscuous plasmids can transfer between bacteria from different species and are maintained stably within them. Plasmids of the IncP, W and Q incompatibility groups can enter and persist in almost all Gram-negative bacteria (reviewed by Thomas and Smith, 1987). The transfer range of promiscuous plasmids often exceeds the range of species in which they can be stably maintained. Broad host range plasmids have been shown to mobilize DNA from Gram-negative bacteria to a number of Gram-positive species and to yeast, demonstrating the possibility of horizontal gene transfer between genera and kingdoms. Even if a plasmid cannot replicate after transfer to an exotic species, it will persist for some time in the recipient's gene pool where there is opportunity for recombination with a host replicon. Recombinational rearrangements will be stimulated in organisms such as gonococci species which cleave circular DNA during uptake (Biswas *et al.*, 1985) or even where incoming DNA is cleaved by host restriction enzymes.

The best studied of the promiscuous plasmids belong to the IncP group; in particular the IncPα plasmid known variously as RP1, RP4, RK2, R18 and R68 has received attention in many laboratories. The presence of a complex series of co-regulated operons, suggests that there is a sophisticated and co-ordinated system of control over plasmid replication, maintenance and transfer functions (Guiney and Lanka, 1989; Thomas and Helinski, 1989). The plasmids are self-transmissible at high frequency on a solid substrate but, character-istically for plasmids which encode a rigid sex pilus, mating in liquid is inefficient. Transfer functions are localized in clusters: Tra1 (13 kb) contains the *traF, -G, -H, -I, -J* and *-K* genes which are involved in DNA transfer. Tra2 (11.2 kb) contains 12 genes designated *trbA–L* (Lessl *et al.*, 1993; Lessl *et al.*, 1992) which are required primarily for mating pair formation.

Despite the greater host range of IncPα plasmids, the mechanism of transfer is probably analogous to that of F and its relatives; the amount of DNA devoted to *tra* functions is certainly similar. RK2 requires an origin of transfer (*oriT*) and at least eight functions encoded within Tra1 and Tra2. The plasmid can be isolated as a protein–DNA relaxation complex in which the TraH, TraI, TraJ and TraK proteins interact with *oriT*. Treatment of this relaxosome with a detergent introduces a specific nick at *oriT* and the 5′ terminus of the nick is linked covalently to Tyr^{22} of the TraI (relaxase) protein (Pansegrau *et al.*, 1993). Presumably this mimics the event which initiates initiating rolling circle DNA replication and plasmid transfer *in vivo* (Waters and Guiney, 1993).

5.4.6 Mobilization

Mobilization is a process by which plasmids which are too small to encode a full set of genes for conjugation can achieve transfer by 'borrowing' the gene products of a self-transmissible plasmid. Examples are *E. coli* plasmids ColE1 (6.6 kb) and RSF1010 (8.9 kb). The role of the conjugative plasmid is to provide pilus-mediated cell-to-cell contact, formation of a conjugation pore and related morphological functions. The mobilized plasmid must possess an active *oriT* (origin of transfer) and a small set of plasmid-specific *mob* genes. Mob proteins bound to *oriT* are responsible for nicking the mobilizable plasmid and initiating strand transfer. In ColE1 and related plasmids ColK and ColA, genetic analysis of the 2.1 kb *mob* region has revealed four essential genes (Boyd *et al.*, 1989). In RSF1010, the transfer origin and three *mob* genes are contained within a 1.8 kb region (Derbyshire *et al.*, 1987). RSF1010 is a member of the broad-host-range IncQ incompatibility group and it provides an extreme example of how a small set of *mob* genes can create the potential for plasmid-mediated gene transfer across species barriers. Assisted by the products of the Ti plasmid *vir* genes, RSF1010 will transfer from *Agrobacterium tumefaciens* into plant cells (Buchanan-Wollaston *et al.*, 1987).

5.4.7 Universal plasmid transfer by conduction

Plasmids which are non-transmissible and non-mobilizable may still be transferred by conduction. This involves a physical association with a transmissible or mobilizable plasmid which may arise through the movement of replicative transposons, or rare plasmid fusions mediated by homologous recombination or plasmid multimer resolution systems. The non-mobilizable plasmid is a passive participant, transferred as a co-integrate with its mobile partner which may resolve into two plasmids in the recipient. Conduction is potentially of great importance in plasmid ecology because, if sufficient opportunities for conduction exist, the concept of a non-transmissible plasmid is meaningless.

The ubiquity of multimer resolution systems provides a means for the formation of transient intermediates between mobilizable and non-mobilizable plasmids and an opportunity for conduction. Recombination occurs efficiently between multimer resolution sites from related plasmids (e.g. ColE1 *cer* and ColK *ckr*; Summers *et al.*, 1985) and although the reaction is normally restricted to resolution, plasmid fusions occur at low frequency. When a donor containing a self-transmissible plasmid, a mobilizable *cer*⁺ plasmid and a non-mobilizable *cer*⁺ plasmid was mated with a recipient defective in *cer*-mediated dimer resolution, cointegrates between the mobilizable and non-mobilizable plasmids were 'trapped' in the recipient at 10^{-4} of the frequency at which the mobilizable plasmid was transferred (H. Chatwin, H. Kaur and D.K. Summers, unpublished data).

5.4.8 Conduction of the *E. coli* chromosome in Hfr strains

The best-known example of conduction occurs in HFr strains of *E. coli* where plasmid F has fused with the *E. coli* chromosome by recombination between mobile elements common to both. Despite the huge difference in size between F and the chromosome, transfer remains unidirectional from the plasmid *oriT*. This gives rise to the useful property of HFr strains that the relative distance of a gene from the origin of transfer can be deduced from its time of entry into the recipient. A set of HFr strains with F integrated throughout the chromosome is available for use in gene mapping by the interrupted mating technique (Brooks Low, 1987).

5.5 Conjugation between Gram-positive organisms

Transmissible plasmids have been reported in numerous Gram-positive genera including *Streptococcus*, *Staphylococcus*, *Bacillus*, *Clostridium*, *Streptomyces* and *Nocardia*. The early stages of the process are not well-characterized but pili do not appear to be involved in initiating conjugation.

5.5.1 Enterococcal plasmid transfer

A feature of conjugation so far unique to the enterococci is the involvement of pheromones or clumping-inducing agents (reviewed by Clewell, 1993). Pheromones are hydrophobic polypeptides of 7–8 amino acids produced by potential recipient cells. Each cell may produce multiple chemical signals, inviting attention from donors containing a variety of conjugative plasmids; plasmid-free cells secrete at least five. The effect of the pheromone on the donor is to induce synthesis of a proteinaceous adhesin which stimulates cell clumping and allows conjugation to proceed. Once a cell acquires a particular plasmid, it stops secreting the corresponding pheromone. Furthermore, the plasmid encodes a protein which behaves as a competitive inhibitor of pheromone action which may prevent self-induction in the event of incomplete shutdown of pheromone synthesis.

pAD1 from *Enterococcus* (formerly *Streptococcus*) *faecalis* is one of the best-studied of pheromone-induced plasmids. At least half of its genome (30 kb) is devoted to conjugation and related functions whose expression is induced by the pheromone cAD1. Tanimoto and Clewell (1993) have proposed a model for the pheromone regulation of the mating response based on transcriptional attenuation.

Some conjugative plasmids in enterococci do not use pheromones. These plasmids transfer poorly in broth unless present in a cell with a pheromone-responsive plasmid. In many ways this is analogous to the effect of self-transmissible plasmids on mobilizable elements in *E. coli*.

5.5.2 Conjugal transfer among the *Streptomyces*

Plasmids are widespread among *Streptomyces* species and conjugation is the most closely studied mechanism of gene transfer in this genus (Rafii and Crawford, 1989; Hopwood and Kieser, 1993). Conjugative plasmids carry genes which promote their transfer to other *Streptomyces* species and, in some cases, support the mobilization of chromosomal DNA and non-conjugative plasmids. In contrast to the linear transfer of chromosomal genes by F-like plasmids, chromosomal markers in the *Streptomyces* are mobilized in an apparently random manner.

Pock formation is a curious and characteristic feature of *Streptomyces* conjugative plasmids which has greatly simplified their detection. Pocks are circular areas of retarded growth which appear around plasmid-containing colonies growing on a lawn of plasmid-free cells. The phenomenon is also known as lethal zygosis on the hypothesis (by analogy with observations of $F^+ \times F^-$ matings) that some recipients may be killed when they encounter multiple donors around the periphery of the colony.

The conjugative multicopy plasmid pIJ101 has been studied in detail. At only 8.8 kb it is significantly smaller than the smallest conjugative plasmid from Gram-negative organisms but is nevertheless remarkably efficient. It can transfer to plasmid-free bacteria with almost 100% efficiency and promotes recombination between the chromosomes of mating bacteria, yielding up to 1% recombinant cells (Kieser *et al.*, 1982). The fact that less than half of the pIJ101 genome is devoted to transfer and fertility implies a much less elaborate mechanism than has been described for F-like plasmids. A single transfer gene (*tra*) and its associated repressor (*korA*) are essential for conjugation. Two further genes (*spdA* and *spdB*) enhance the frequency of transfer. Transfer also requires the *cis*-acting *clt* locus which, if deleted, reduces pIJ101 transfer over 1000-fold (Pettis and Cohen, 1994). A different genetic organization has been reported for the large (31.4 kb), low copy number plasmid SCP2 in which five transfer-related genes are clustered in a region of 9 kb (Brolle *et al.*, 1993). Despite their differences, the transfer regions of pIJ101 and SCP2 both confer derepressed plasmid transfer and pock formation.

5.5.3 Conjugation among the staphylococci

Self-transmissible antibiotic resistance plasmids of 38–57 kb are found in the staphylococci (Lyon and Skurray, 1987) and genes encoding conjugation functions occupy about a third (12–15 kb) of the plasmid genome. The mechanism of conjugation in the staphylococci is not well-characterized but the relatively small amount of DNA coding for the necessary functions suggests the process may be simpler than that employed by plasmid F which devotes twice as much DNA to transfer functions.

At high cell density staphylococcal plasmids may also be exchanged by 'phage-mediated conjugation' which requires the presence of a high concentration of calcium ions and a prophage in either the donor or recipient. It has been suggested that the prophage enhances cell clumping by increasing the surface adhesiveness. Further evidence that changes in the cell surface facilitate plasmid transfer comes from observations that sub-inhibitory concentrations of β-lactam antibiotics stimulate phage-mediated conjugal transfer of a tetracycline resistance plasmid between 100- and 1000-fold.

5.5.4 T-DNA: *trans*-kingdom conjugation

An astonishing manifestation of the versatility of plasmids as agents of programmed gene transfer is provided by the large extrachromosomal elements found in pathogenic species of *Agrobacterium*. *A. tumefaciens* infects wound tissue in dicotyledons and causes the formation of crown gall tumours while *A. rhizogenes* stimulates root proliferation and is the causative agent of hairy root disease. In both cases pathogenicity is associated with the presence of a large self-transmissible plasmid, known as Ti (tumour inducing) or Ri (root-inducing). The Ti plasmid has been the subject of extensive study and is the paradigm for the transfer of genetic material between distantly-related species (reviewed by Winans, 1992).

The mechanism of gene transfer mediated by the Ti plasmid is closely analogous to conventional bacterial conjugation. Cell-to-cell contact is an essential prerequisite and is established when *A. tumefaciens* binds to damaged plant cells. A single-stranded segment of plasmid DNA (the T-DNA) whose boundaries are defined by imperfect 25 bp direct repeats, is mobilized from the donor plasmid into the nucleus of a susceptible plant. After transfer it integrates at random into the plant genome. If the T-segment is replaced with foreign DNA, it will be transported into the plant genome, making *A. tumefaciens* a powerful delivery system for plant genetic manipulation.

Approximately 25 virulence genes (designated *vir* and analogous to *tra* and *mob* in conventional plasmid systems), comprise six or seven operons in a 35 kb region of the Ti plasmid not transferred to the plant cell. The gene products promote the establishment of cell-to-cell contact and mediate transfer of the T-DNA. Gene transfer is initiated when the VirD2 endonuclease introduces single-stranded nicks at the direct repeats which define the termini of the transferred region (analogous to nicking at *oriT* of plasmid F; Waters and Guiney, 1993), and becomes attached covalently to the 5′ end of each nick. The single-stranded DNA molecule thus generated is transferred unidirectionally, as part of a complex with Vir proteins which protect and direct it, to the plant cell nucleus. Entry into the nucleus may be facilitated by the VirD2 protein which appears to contain a nuclear localization signal.

Following the integration of T-DNA into the plant genome, genes encoding three biosynthetic pathways are expressed. Two pathways direct synthesis of the plant growth stimulators auxin and cytokinin which release the cell from normal growth constraints and promote tumour formation. The third directs the production of an opine which, as well as providing a rich source of carbon and nitrogen for the pathogenic bacterium, has a role in the regulation of plasmid transmission. The opine acts as an inducer of conjugal transfer genes, leading to derepressed transmission of the Ti plasmid. This element of positive feedback results in an epidemic of plasmid transfer.

5.5.5 Non-plasmid conjugation systems

Gene transfer by conjugation is not invariably plasmid-mediated. A family of transposable elements including Tn*916* (a 16.4 kb tetracycline resistance transposon from *Enterococcus faecalis*) and Tn*1545* (25.3 kb from *Strep. pneumoniae*) are capable of conjugal transfer at frequency of 10^{-6}–10^{-8} per generation in the absence of a plasmid in either donor or recipient. A class of larger (> 50 kb) elements with similar properties includes Tn*3701*, Tn*3951* and Tn*5253*. These have been less well studied and it has been proposed that they may be genetically distinct from Tn*916* (Vijayakumar and Ayalew, 1993).

In common with conventional transposons, the Tn*916*-like elements are flanked by imperfect terminal inverted repeats (20–26 bp) but they provide an exception to the rule that the target sequence is duplicated at the site of insertion. Known as conjugative transposons (reviewed by Scott, 1992), they are extremely promiscuous; there is no requirement for donor and recipient to belong to the same species or genus and the transferred DNA appears to be immune from host-mediated restriction. They provide a general mechanism for the transmission of antibiotic resistance determinants among pathogenic bacteria and are therefore of considerable medical significance (see section 6.2.7).

Conjugative transposition proceeds through a non-replicative, circular DNA intermediate formed by site-specific excision of the element from its parent replicon (Fig. 5.3). The enzymology of transposition is closely related to the integration and excision of lambdoid bacteriophages; excision requires the joint action of an excisionase (Xis-Tn) and an integrase (Int-Tn) while Int-Tn alone is sufficient to mediate subsequent integration (Poyart-Salmeron *et al.*, 1990; Scott *et al.*, 1994). Little is known about the transfer of the intermediate from donor to recipient although it has been suggested that cell fusion leads to the formation of a transient zygote. Once in the donor there is no strong specificity for a target sequence but there is a preference for A : T-rich regions and for sites showing homology with the ends of the element.

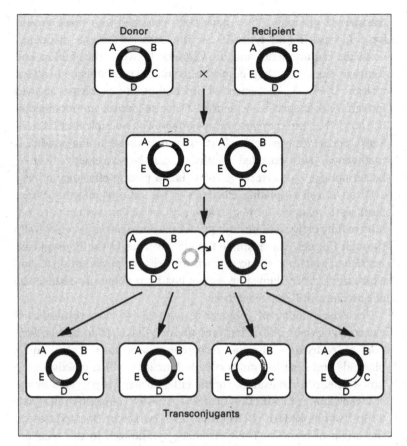

Fig. 5.3 Conjugative transposition. The stippled region in the donor chromosome represents the transposon. Physical contact between donor and recipient cells triggers excision of the transposon to form a circular, non-replicating intermediate. The transposon is transferred to the recipient where it integrates into the chromosome. From Scott (1992).

5.6 Limitations on gene transfer

The multiplicity of mechanisms for gene transfer among the prokaryotes seems to cast doubt on the usefulness of the species concept in prokaryotes. However, these mechanisms operate at low frequencies even under ideal conditions. In the natural environment, a particular pair of species may rarely come into close proximity and, even gene exchange occurs, there is no guarantee that the transferred material will remain intact or be expressed in the new host.

5.6.1 Restriction and modification of foreign DNA

Restriction-modification systems (reviewed by Bickle and Krüger, 1993) present a major threat to the physical integrity of newly transferred plasmids. They have been classified into three groups (type I, type II and type III) on the basis of subunit structure, co-factor requirements and the distance between recognition and cleavage sites, but they employ a common mechanism to attack foreign DNA. A host-encoded restriction endonuclease cleaves foreign DNA at specific sequences. These sequences are modified in the host DNA and therefore escape attack. For example Sau3 A from *Staph. aureus* cleaves at 5'-GATC-3' unless the thymine residue is methylated. Host-encoded restriction can be extremely effective; bacteriophage λ grown on *E. coli* C has a plating efficiency of only 10^{-4} on *E. coli* K (plating efficiency is the ratio of infecting phage particles to plaques formed). However, restriction systems are not always fully active and incoming DNA may sometimes be able to slip through the net. For example the effectiveness of the *P. aeruginosa* restriction system is reduced when the cells are grown at 43°C and exposure to DNA damaging agents has been shown to reduce the effectiveness of *E. coli*'s defences.

The susceptibility of conjugative plasmids to host restriction is extremely variable. Selection pressure may have led to the eradication of restriction enzyme recognition sites from promiscuous plasmids and some plasmids encode functions which inactivate the host's defences. An example of the latter strategy is provided by the *ard* (alleviation of restriction of DNA) genes of conjugative plasmids ColIb-P9 and pKM101 (Delver *et al.*, 1991) which are located close to the origin of transfer. Their products, synthesized in the recipient very early in conjugation, afford protection against type I restriction-modification systems.

5.6.2 Gene expression in exotic hosts

Plasmid persistence in novel host species requires not only the physical transfer of DNA but also the expression of plasmid-encoded information. Essential genes include those for plasmid replication and maintenance, and proliferation is aided by genes which confer phenotypes such as drug resistance or pathogenicity. There is evidence that promiscuous plasmids encode functions which maximize the likelihood of establishment in a new host. These include the Sog proteins (DNA primases), which are synthesized in the donor and injected into the recipient where they initiate complementary strand synthesis independently of the host cell primases (Wilkins *et al.*, 1991). Another class of protein (PsiB) whose role is less well understood appears to inhibit the induction of the bacterial SOS response during conjugation.

Problems with gene expression in exotic hosts may result from misreading of signals during initiation, elongation or termination of transcription or translation. Even when produced, transcripts and proteins may be unstable or may not be processed correctly. When the *Pseudomonas* TOL plasmid, which enables its host to use toluene and xylenes as sole carbon and energy sources, was introduced into *E. coli* it was able to replicate and was conjugationally proficient but failed to alter the catabolic phenotype of the host (Benson and Shapiro, 1978).

5.6.3 Have plasmids avoided the perils of Babel?

Our knowledge of control signals is limited to a narrow range of species but there is reason to believe that cross-recognition of signals (albeit rather inefficient) may occur between widely divergent organisms. There has been evolutionary conservation of RNA

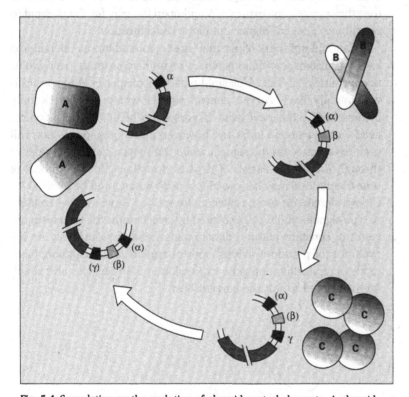

Fig. 5.4 Speculation on the evolution of plasmid control elements. A plasmid crossing from species A to B may find that a control sequence (α) is recognized inefficiently by the new host. If the plasmid persists for long enough in its new host, natural selection may produce a functional control element (β) either *de novo* or by modification of α. Further wandering brings the plasmid into species C where yet another control sequence (γ) is required. Finally a return to species A demands the renovation of the original sequence. In this way a gene on a well-travelled plasmid may accumulate a complex control region containing the remnants of sequences recognized in many species.

polymerases in a wide range of bacteria and at least some promoters in *E. coli*, *Pseudomonas* species, *Strep. pneumoniae* and *B. subtilis* appear to be closely related. Even in the *Corynebacteria* where classical *E. coli* promoters are not obviously apparent, some expression of *E coli*-derived genes (including kanamycin resistance) has been observed. This provides hope that even after substantial interspecific jumps, plasmids may still be able to manage a limited degree of gene expression.

Poor expression of plasmid genes need not provide an insurmountable barrier to plasmid spread. Inefficient post-conjugal gene expression conferring resistance to low concentrations of an antibiotic could be sufficient to increase fitness and maintain the plasmid in a new host species on the edges of an antibiotic-containing environment. There would then be strong selection for mutations affecting replication, transcription and translation, which increase the efficiency of gene expression and open the way for the plasmid to spread into areas of higher antibiotic concentration.

We should not think of plasmid–host combinations as unchanging but as transient associations which reshuffle constantly over evolutionary time (Fig. 5.4). A plasmid passed from species A to species B will initially have A-type control signals which may work only inefficiently in the new host. If possession of the plasmid confers even a small increase in fitness, however, mutation and selection will produce a set of B-type signals while the A-type signals may decay through the accumulation of neutral mutations. In the fullness of time the plasmid may be passed to a third host (C) and be selected for C-type signals or even returned to species A when the residual A-type signals would be renovated by selection. In the long term the result of constant plasmid dissemination and recycling may be the evolution of pseudo-universal control regions which support basal levels of replication and gene expression in many species and which are fine-tuned to suit the current host.

6: The Clinical and Veterinary Importance of Plasmids

6.1 Plasmids and the spread of antibiotic resistance

Plasmids are important causative agents of gene flow within and between bacterial species. They provide a reservoir of genetic information which compensates for the restricted genome size of prokaryotes, and provides the raw material for astonishingly rapid adaptation of bacterial populations faced with changing environments. One of the most dramatic and well-documented examples of plasmid-driven evolution is the spread of bacterial antibiotic resistance since the Second World War.

6.1.1 Early observations of drug resistance

Drug resistance in microorganisms is not a new phenomenon. As early as 1907 it was reported that *Trypanosoma brucei* became resistant to the effect of *para*-rosaniline after repeated exposure to the drug (Ehrlich, 1907). In bacteria, Morgenroth and Kaufmann (1912) reported that pneumococci could acquire resistance to optochin (ethyldihydrocupreine hydrochloride).

Following the introduction of chemotheraputic agents and antibiotics there were early reports of bacterial resistance to sulphanilamide (Maclean *et al.*, 1939) and penicillin (Abraham *et al.*, 1941). Resistant organisms were found at low frequency in untreated populations (Demerec, 1948). Typically resistance was due to spontaneous mutations which mapped to the bacterial chromosome and affected the site of action of the antibiotic. For example, variants of *Escherichia coli* resistant to high levels of streptomycin were found to be mutant at a single locus (known originally as *strA* but later renamed *rpsL*) and produced an altered 30S ribosomal subunit which was not sensitive to the action of streptomycin (Flaks *et al.*, 1966).

Early reports of antibiotic resistant bacteria failed to arouse the interest of clinicians. If this seems surprising in retrospect we should remember that the contemporary view of bacterial evolution was grounded firmly in concepts of mutation and selection. Phenotypic leaping through the acquisition of new DNA was unknown. It was assumed that each drug resistance resulted from an independent mutational event, so the chance of an organism becoming simultaneously resistant to more than one drug was very small indeed. Multiple resistance could arise if resistance to a new drug arose in an organism already resistant to another. In this way multiply resistant strains were expected to arise only in environments where antibacterial agents were common. The danger of extrapolating data from the laboratory to the field was amply illustrated by the events of the next few decades. By the 1960s it was apparent that the acquisition of drug resistance in nature involved not the mutational alteration of existing genes but the acquisition of extra DNA and that by this mechanism bacteria could simultaneously acquire resistance to several antibiotics.

As unexpected as the infectious spread of multiple drug resistance was the discovery that resistance in clinical isolates and laboratory strains is underpinned by different mechanisms. For example, although penicillin resistance in clinical isolates of staphylococci typically involves hydrolysis of the antibiotic by β-lactamase, this is never seen in laboratory isolates. Similarly, resistance to chloramphenicol often involves acetylation of the drug in clinical isolates but not in strains selected in the laboratory. In general, resistance in laboratory strains is achieved by modification of the target whereas in clinical isolates specialized proteins modify or remove the antibiotic. The failure to isolate the latter class in the laboratory is a reflection of the much smaller, relatively uniform populations involved. Outside the laboratory selection acts upon heterogeneous consortia of microorganisms where opportunities exist for gene exchange between species. It is in these environments that drug resistance plasmids (R factors) come into their own.

6.1.2 Japanese studies

Although evidence that antibiotic resistance was frequently plasmid-borne came from a variety of sources, much of the seminal work was undertaken by Japanese workers studying the epidemiology of bacterial dysentery. Because the majority of this work was published only in Japanese language journals, it was not widely appreciated in the West until the publication of a review in English by Watanabe (1963). Fifty-five out of 148 references in Watanabe's review were to articles published in Japanese. Recognizing the long-term significance of the principles uncovered in the Japanese studies Watanabe commented:

> Initially, the problem of multiple drug resistance received
> attention because of its medical importance, but more recently
> much effort has been devoted to genetic studies from which
> the episomal nature of the responsible factors is emerging as
> one of the most interesting problems.

6.1.3 The rise of multiply resistant bacteria

A detailed account of early work on the epidemiology of drug
resistance can be found in Mitsuhashi (1971). Despite improvements
in sanitary conditions in Japan between 1952 and 1967 the incidence
of bacterial dysentery remained high. Sulphanilamide derivatives had
been used against *Shigella* since 1945 but these were effective for only
about five years before resistant strains appeared. By the early 1950s
resistant strains accounted for 80–90% of clinical isolates. It was
around this time that the availability of a series of antibiotics
including streptomycin, tetracycline and chloramphenicol promised
a solution to the problem. In reality, however, they provided only a
temporary respite before the advent of multiply drug resistant strains.

Multiply resistant *Shigella* accounted for less than 0.02% of clinical
isolates in 1955 but their rise was astonishingly rapid. By 1967 over
74% of isolates were resistant to more than one antibiotic (Fig. 6.1).
The first indication of things to come was in 1956 when members of
a Japanese grocer's family who had been treated with chloramphenicol alone, were found to be carriers of a strain of *Shigella* resistant to
four antibiotics (streptomycin, tetracycline, chloramphenicol and
sulphonamide). Even more alarming was the discovery that *E. coli*
isolates from these individuals were resistant to the same combination of drugs, implying that resistance was transmissible between
species. The rise of antibiotic resistant strains was not confined to
Japan; in the United Kingdom, multiple resistance first appeared in

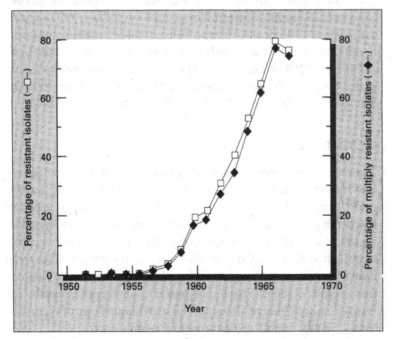

Fig. 6.1 There was a
dramatic increase in
antibiotic resistant *Shigella*
isolates during the decade
after 1955. Almost all of
these isolates displayed
multiple resistance.

1958 and by 1970, 70% of *S. sonnei* strains were resistant to three or more drugs (Datta, 1984).

6.1.4 Transfer of antibiotic resistance

Transfer of drug resistance *in vivo* and *in vitro* was soon demonstrated in the laboratory. In one of the first of a multiplicity of studies, Kasuya (1964) described the transfer of multiple resistance from *S. flexneri* to *E. coli* and *Klebsiella pneumoniae* in the intestines of mice and in the absence of antibiotics. Reed *et al.* (1969) and Jones and Curtiss (1970) observed transfer between strains of *E. coli* under similar conditions. In studies on the human gut, Anderson *et al.* (1973) were only able to demonstrate transfer under antibiotic selection which favoured the growth of resistant cells. Subsequently, however, Petrocheilou *et al.* (1976) reported transfer between strains of *E. coli* in the absence of selection, although it was 204 days before a novel resistant strain was detected.

Plasmid transfer between bacterial species outside the laboratory can have serious clinical implications. In 1983, an outbreak of multiply resistant *S. flexneri* infections on a Hopi Indian reservation was traced to an individual who had undergone long-term treatment with trimethoprim–sulfamethoxazole for urinary tract infection by *E. coli*. Epidemiological studies showed that the *S. flexneri* strain had acquired its drug resistance by the transfer of a plasmid from *E. coli* (Tauxe *et al.*, 1989).

Resistant strains are not found exclusively in individuals undergoing intensive antibiotic treatment. A Japanese study in 1961 showed that 1.4% of 1145 healthy human subjects carried multiply resistant *E. coli* (Mitsuhashi *et al.*, 1961) and more than 20 years later Levy (1986) reported that the majority of faecal samples from healthy people in the Boston area contained substantial numbers of drug resistant bacteria, with about half of these displaying multiple resistance.

6.1.5 R factors

The rapid appearance of multiply resistant strains and evidence for the transfer of resistance between species stimulated interest in the underlying mechanism. Resistance was found to be transmitted independently of the chromosome (Mitsuhashi *et al.*, 1960a) and to be lost from cultures of *Shigella* or *E. coli* during storage or upon treatment with acriflavin (Mitsuhashi *et al.*, 1960b). Parallels were drawn between the transfer of resistance and the cell-to-cell transmission of the F factor by conjugation. The agent responsible for transfer of antibiotic resistance was known initially as *rtf* or *rta* but from 1962 the term R factor became generally accepted.

6.2 The evolution of multiple antibiotic resistance

In this section our attention is focussed on the role of plasmids in the spread of antibiotic resistance. It would be wrong, however, to imply that multiple drug resistance is invariably plasmid-borne. A prime exception is the notorious methicillin resistant *Staphylococcus aureus* (MRSA) which appeared in Europe after the introduction of this semi-synthetic penicillin in the face of resistance to more conventional β-lactam antibiotics. In the 1980s the strain began to cause problems in the United States. As a result of the accumulation of further resistance determinants by transposition and site-specific integration events, the chromosomes of many MRSA strains now encode resistance to additional antibiotics including erythromycin, fusidic acid, tetracycline, streptomycin and sulphonamides

The biology of R factors can be considered at three levels. First, how do plasmids gain and lose resistance genes? Second, how do these plasmids spread among bacterial strains and third, by what routes do resistant organisms spread through human and animal populations? This section concentrates on the first of these with particular reference to the roles of transposons and integrons. The means by which plasmids are transmitted between bacteria was the subject of chapter 5 while the mechanisms by which antibiotic resistant organisms spread are beyond the scope of this book.

6.2.1 Plasmid evolution is modular

Plasmid evolution is a saltatory phenomenon, proceeding by the acquisition and loss of functional modules rather than the gradual processes of mutation and selection (see Amabile-Cuevas and Chicurel, 1992 for an overview of the role of plasmids in bacterial evolution). The processes which reshuffle blocks of sequence between and within antibiotic resistance plasmids generate the structural and phenotypic variation which is essential for their evolution. These processes include transposition and homologous, illegitimate and site-specific recombination.

Evidence of modular evolution has been reported for the conjugative plasmids of the RepIIA family of *E. coli* (Lopez *et al.*, 1991) and multicopy plasmids of Gram-positives (Projan and Moghazeh, 1991). R factors from clinical isolates often show a high degree of structural variation; a property which presumably underpins the ability of bacteria to respond with alarming rapidity to the use of novel combinations of antibiotics. An illustration of structural heterogeneity and the role of transposable elements was revealed in a study of the multiresistance plasmid pBP16 and its derivatives. A series of rearrangements including inversion, deletion, replicon fusion and dissociation was traced to the presence of multiple copies of the insertion sequence IS*60* (Nies *et al.*, 1986).

6.2.2 Transposons and antibiotic resistance

Antibiotic resistance genes often form an integral part of transposable elements which are capable of movement from one plasmid to another, or between plasmids and the chromosome. Although transposons are unable to transfer resistance between cells, they can effectively mobilize genes by moving them to a transmissible plasmid. Since promiscuous plasmids have host ranges which include bacteria from different genera the opportunity exists for the dissemination of transposon-borne resistances among a very wide variety of organisms. Even if the conjugative plasmid is unable to replicate its new host, resistance genes may be rescued by movement of the transposon to a native replicon or by recombination with a plasmid or the chromosome in the recipient. A study of the role of transposons and broad-host-range plasmids in the dissemination of ampicillin resistance among *Neisseria* and *Haemophilus* species has been described by Saunders *et al.* (1986).

6.2.3 Class I transposons

Transposons in bacteria were discovered through their role in the spread of antibiotic resistance. In Gram-negative bacteria antibiotic resistance transposons are divided into two groups on the basis of

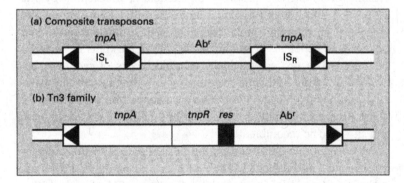

Fig. 6.2 Bacterial transposable elements. (a) Class I or composite transposons consist of insertion sequences (IS) on either side of an antibiotic resistance determinant (Ab^r). Each IS contains a transposase gene (*tnpA*) flanked by inverted repeats to which the transposase binds. (b) Members of the Tn*3* family have a more complex structure. They encode a transposase (*tnpA*) and a resolvase (*tnpR*) together with accessory functions (Ab^r) which may include resistance to antibiotics and heavy metals. The resolvase catalyses site specific recombination between *res* sites to process replicon fusions (cointegrates) resulting from inter-molecular transposition.

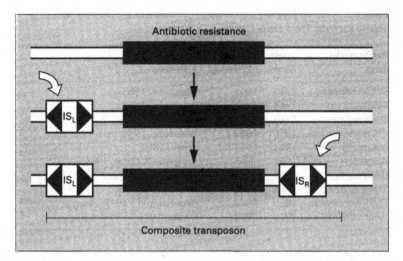

Fig.6.3 Genesis of a composite transposon. An antibiotic resistance gene becomes part of a composite transposon when it is flanked by insertion sequences.

structure and their mechanism of transposition (Grindley and Reed, 1985). Class I transposons (Fig. 6.2a) have a sandwich or composite structure with resistance genes flanked by copies of an insertion sequence (IS). Insertion sequences range from 750 to 1600 bp and contain the information for synthesis of a specific transposase. In some elements such as Tn*10*, mutations have destroyed the transposase gene in one of the IS elements.

It is easy to imagine the evolutionary origin of composite transposons (Fig. 6.3). Insertion sequences are capable of independent transposition and any chromosomal sequence flanked by IS copies becomes, by default, a composite transposon which can move as a single unit. In environments where organisms are exposed to high levels of antibiotics, natural selection will assist the proliferation of transposons which incorporate genes for drug resistance. This appears to have happened for the class I transposons Tn*9* and Tn*10* which were responsible for resistance to chloramphenicol and tetracyclines reported in the late 1950s (Table 6.1).

The relatively high frequency with which composite antibiotic resistance transposons arise is indicated by a comparison of Tn*5* and Tn*903*. In both cases a gene encoding kanamycin resistance is flanked by a pair of IS elements (IS*5* and IS*903*, respectively). Sequence comparison shows that the resistance genes in the two transposons are related but that the flanking IS elements are not, suggesting that the same resistance gene was acquired independently by the two transposons (Grinsted, 1986).

6.2.4 Class II transposons

Tn*3* is the paradigm for the widespread and rather diverse group of class II transposons known as the Tn*3* family (Fig. 6.2b). These

Table 6.1 A chronology of plasmid-borne resistance to antibacterial agents in enteric bacteria. From Datta (1984).

Antibacterial drug	In use since:	R-plasmid discovered	Reference
Sulphonamides	1936	1959	Watanabe, 1963
Streptomycin	1948	1959	Watanabe, 1963
Tetracyclines	1949	1959	Watanabe, 1963
Chloramphenicol	1950	1959	Watanabe, 1963
Neomycin	1954	1963	Lebek, 1963
Ampicillin	1962	1965	Anderson & Datta, 1965
Gentamicin	1964	1972	Witchitz & Chabbert, 1972
Trimethoprim	1968	1972	Fleming *et al.*, 1972
Phosphonomycin	1969	1977	Baquero *et al.*, 1977; Perea *et al.*, 1977
Amikacin	1974	1974	Jacoby, 1974

elements are flanked by short inverted repeats (38 bp for Tn3) and contain genes encoding the transpose and resolvase enzymes, together with resistance to one or more antibiotics. The resolvase is not required for transposition but dissects cointegrate molecules in which the donor and target replicons are fused following intermolecular transposition.

While the genetic organization of class I composite transposons suggests an obvious pathway for their evolution, the origin of class II transposons is less clear. Comparative studies of transposase genes suggest the family should be divided into two groups typified by Tn21 and Tn3. Within each group, transposase genes share about 70% DNA sequence identity and their products cross-complement (although not always with full efficiency). Between the groups, transposases fail to complement and show no more than 50% DNA sequence identity, suggesting that Tn3 and Tn21 diverged long ago.

6.2.5 Transposons, integrons and the acquisition of resistance genes

Most multiple drug resistant isolates of Gram-negative bacteria contain a member of the Tn21 group of transposons and it is often among these elements that resistance to novel antibiotics first appears. Tn21 carries genes which confer resistance to sulphonamides (*sul*) and spectinomycin (*aadA*). Related elements (Table 6.2) contain additional resistance genes located on either side of *aadA* and it was suggested by Schmidt (1984) that *aadA* might be flanked by sites which act as hot-spots for the acquisition of additional resistance genes. Analogous acquisition systems or integrons (reviewed by Hall and Stokes, 1993) have been found in Tn7, the IncW plasmids R388 and pSa, the IncN plasmid R46 and in various pLMO

Table 6.2 Antibiotic Resistances conferred by the Tn*21* sub-family.

Element	Genotype	Resistance	Reference
Tn*21*	*sul*	Sulphonamides	de la Cruz and Grinsted, 1982
	aadA	Streptomycin	
Tn*501*	*mer*	Mercury	Bennett *et al.*, 1978
Tn*3926*	*mer*	Mercury	Lett *et al.*, 1985
Tn*1721*	*tet*	Tetracycline	Schmitt *et al.*, 1979
Tn*2424*	*aacA*	Amikacin	Meyer *et al.*, 1983
	cat	Chloramphenicol	
Tn*2603*	*bla* OXA-1	Ampicillin	Yamamoto *et al.*, 1981
Tn*2410*	*bla* OXA-2	Ampicillin	Kratz *et al.*, 1983
Tn*4000*	*aadB*	Gentamicin	Schmidt, 1984
Tn*2501*		Cryptic	Michiels and Cornelis, 1984

plasmids which encode resistance to trimethoprim (see Grinsted *et al.*, 1990 and references therein).

The functional organization of integrons (Stokes and Hall, 1989; Hall and Stokes, 1993) has been deduced by comparative studies of Tn*21*, and related transposons. Integrons can be divided into three discrete regions: a 5' common segment, a central variable region and a 3' common segment (Fig. 6.4). The 3' common segment contains *sul* (resistance to sulphonamides) and two additional open reading frames: *qacEΔl* (a defective quaternary ammonium compound

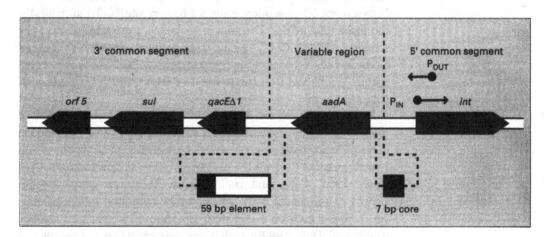

Fig. 6.4 Genetic organization of integrons. The illustration shows the region containing the streptomycin (*aadA*) and sulphonamide (*sul*) resistance determinants of Tn*21*. The streptomycin resistance determinant is contained in a central variable segment. To the right lies a 59 bp imperfect inverted repeat (heavily outlined box) which is partially conserved among many integrons. At the left end of the 59 bp sequence is a 7 bp 'core' (shaded box) which also appears at the boundary between the variable region and the 5' common segment. The 5' common segment contains the *int* gene whose product is required for the integration of DNA into the variable region. Promoters P_{IN} and P_{OUT} direct transcription of *int* and the variable region genes, respectively.

exporter) and ORF-5 of unknown function. The variable region contains one or more resistance genes (e.g. *aadA* in Tn21). In most cases the variable region genes lack their own promoters and are transcribed as an operon from P_{OUT} which lies at the inner boundary of the 5′ common segment. Thus integrons act as natural expression vectors for insert genes. At the distal end of each resistance gene cassette in the variable region is a 59 bp sequence which displays imperfect dyad symmetry. This sequence is poorly conserved between different systems but it is essential for the acquisition of additional DNA (Martinez and de la Cruz, 1990).

6.2.6 Integron evolution

Within the 5′ common segment of the integron lies the *int* gene which encodes a recombinase required for the insertion of new DNA cassettes in the variable region (Martinez and de la Cruz, 1990). It is transcribed in the opposite direction from the genes in the variable region and sequence analysis suggests that its product belongs to the Int family of site-specific recombinases (see Argos *et al.*, 1986 for a review of these enzymes). Martinez and de la Cruz (1988) observed the Int-dependent fusion of plasmid R388 with a replicon containing Tn21. They concluded that fusion had occurred either by recombination between the 59 bp sequences on the two participants or between one 59 bp sequence and a 7 bp core site (Hall *et al.*, 1991) at the inside end of the 5′ common sequence. Core sites contain the sequence GTTRRRY (R, purine; Y, pyrimidine), which also appears as the last seven bases of the 59 bp sequence. Recombination occurs within or adjacent to a conserved GTT triplet which is found in all core sequences.

Further evidence for the role of site-specific recombination in integron evolution has come from studies of Int-dependent deletion of individual resistance genes and the generation of duplications and rearrangements through the insertion of genes at new locations (Hall *et al.*, 1991; Collis and Hall, 1992a). Resistance gene cassettes behave as functional units which can be mobilized independently. Int-mediated excision of a cassette generates a circular intermediate (Collis and Hall, 1992b) which can reintegrate at a new site by recombination with a core sequence (Collis *et al.*, 1993; Fig. 6.5). Integrons containing a single resistance gene are thought to have arisen by the insertion of a circular gene cassette into a pre-integron (containing no resistance genes and no 59 bp repeat). Subsequent integration of gene cassettes involves recombination between 59 bp sequences provided by the integrated gene and the incoming cassette. Consistent with this hypothesis, an ancestral integron (In0) in which the 3′ and 5′ conserved sequences flank a single core site has been identified in plasmid pVS1 of *Pseudomonas aeruginosa* (Bissonnette and Roy, 1992).

Martinez and de la Cruz (1988) showed that Int-mediated recombination can lead to the formation of cointegrates between integron-containing plasmids. Subsequent resolution by recombination between any pair of sites other than those involved in the initial fusion will lead to deletion or rearrangement. Hall and Stokes (1993) suggest that this mechanism is likely to contribute extensively to the

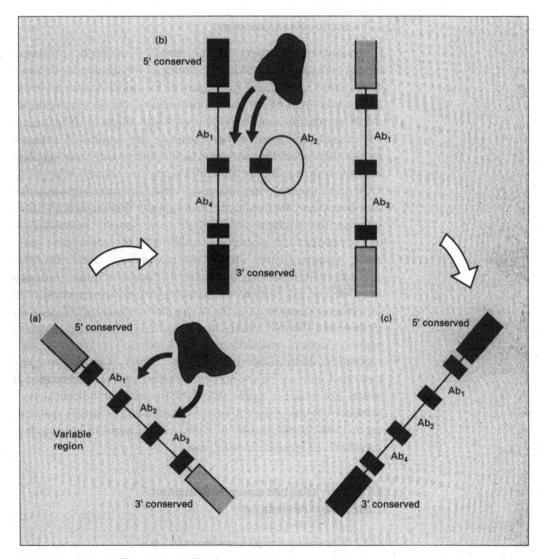

Fig. 6.5 Integron shuffling. (a) Int-mediated recombination between core sequences separating resistance genes Ab_1, Ab_2 and Ab_3 excises a circular cassette carrying Ab_2. (b) The core site on this circular intermediate recombines with a core sequence on an integron carrying determinants Ab_1 and Ab_4. (c) The product of recombination is a new integron carrying Ab_1, Ab_2 and Ab_4.

reshuffling of genes between existing integrons. In addition, less programmed variation may result when integrons insert into sequences which resemble the primary recognition site of the Int recombinase (Recchia *et al.*, 1994; Recchia and Hall, 1995). Finally, the role of RecA-dependent recombination cannot be ignored. Extensive conserved sequences flank the drug resistance genes and recombination between these regions of homology could lead to an exchange of cassettes between integrons.

6.2.7 Conjugative transposons

Of undoubted importance in the spread of antibiotic resistance among pathogenic bacteria is the family of conjugative transposons (Scott, 1992; Clewell and Flannagan, 1993), which includes Tn*916*, Tn*918* and Tn*920* from *Enterococcus faecalis*, Tn*1545* from *Streptococcus pneumoniae* and Tn*919* from *Strep. sanguis*. Unlike more conventional transposons these elements are capable of conjugal transfer between a wide variety of Gram-positive and Gram-negative bacteria in the absence of a plasmid vector (see section 5.5.5).

Terminal inverted repeats disguise these elements as conventional transposons but they generate no target site duplications and they transpose by a unique combination of site-specific recombination and conjugation. In pioneering studies with the tetracycline resistance element Tn*916*, Clewell and co-workers found that, like conventional transposons, this element relocates within its host at a frequency of 10^{-6}. More remarkable was the demonstration of intercellular transfer at ~1% of normal transposition frequency (Franke and Clewell, 1981; Clewell *et al.*, 1991). The first stage of transfer involves excision of the element as a circular, non-replicating intermediate (Scott *et al.*, 1994). This requires the concerted action of a site-specific integrase and an excisionase. It seems likely that a single strand of the circular intermediate is transfered to the recipient where complementary strand synthesis regenerates a double-stranded circle which integrates by integrase-mediated site-specific recombination.

6.3 The clinical consequences of antibiotic resistance

6.3.1 The dawn of a post-antimicrobial era?

The success of the pharmaceutical industry in developing new antibiotics during the last 30 years has encouraged complacency over the threat from microbial drug resistance. In the armoury of clinicians and veterinarians there are now more than 50 penicillins, 70

cephalosporins, 12 tetracyclines, eight aminoglycosides, one mono-
bactam, three carbapenems, nine macrolides, two streptogramins
and three dihydrofolate reductase inhibitors. Despite this, resistant
organisms continue to precipitate hospital ward closures and to cause
death and serious illness throughout the developed word. *Streptococcus pneumoniae*, *Strep. pyogenes* and staphylococci which cause
respiratory and cutaneous infections, and species of enterobacteriaceae and *Pseudomonas* that cause diarrhoea, urinary infections and
sepsis are now resistant to virtually all of the older antibiotics.

The implications of the rapid spread of drug resistance have been
discussed by Cohen (1992). The most pessimistic scenario envisages
a 'post-antimicrobial era' in which clinicians face a return to the
conditions in the 1930s, with hospital wards filled with patients
suffering from pneumonia, meningitis, bacteraemia, typhoid fever,
endocarditis, mastoiditis, syphilis, tuberculosis and rheumatic fever.
Until the introduction of antimicrobial agents, the majority of such
patients died either from the disease or from associated complications.

Although a substantial number of pathogens remain sensitive to
common antibiotics, morbidity and mortality associated with many
diseases is increasing because concern about resistance causes
delays in the application of effective therapy. The application of an
antibiotic to which the pathogen is resistant may have a harmful
effect on the patient by killing sensitive competing organisms and
allowing more rapid proliferation of the pathogen. There are consequences too for the cost of treatment. It has been estimated that the
cost of treating a case of tuberculosis increases from $12 000 when
the causitive organism is antibiotic sensitive to $180 000 if infection
is due to a multiply resistant strain (Fox, 1992).

6.3.2 Strategies for antibiotic development and use

In a discussion of prospects for the prevention and control of
antimicrobial resistance, Cohen (1992) lists several strategies which
might be employed to combat the potential crisis in public health.
The development of new antimicrobial agents to be used on their
own or in combination with existing drugs tends to be taken for
granted but the supply will not be limitless and even after discovery
of a new compound, there is a six or seven year delay before the drug
reaches the market.

In addition to developing new drugs it is important to strive to
minimize the rate at which resistant organisms arise. More selective
use of antibiotics in human and veterinary medicine is essential so
that selection pressures are imposed on the smallest possible population of organisms. The use of broad-spectrum antibiotics gives
clinicians greater confidence when treating poorly characterized

infections but extending selection to a wide variety of co-resident non-pathogenic species may establish a reservoir of resistance genes which are available for transfer to the pathogen. Rapid and accurate diagnosis is essential to allow narrow-spectrum drugs to be used for the minimum time consistent with effective eradication of infection. In this context the importance of preventive medicine should not be overlooked. Improvements in public health practices will reduce the opportunities for infection and the transmission of resistance determinants between species. In agriculture, improved animal husbandry practices can reduce transmission of resistant microorganisms between animals and into the human population.

6.4 Mechanisms of antibiotic resistance

Plasmid-mediated resistance is most often provided by enzymes which modify the antibiotic. For penicillin derivatives this involves enzymatic cleavage of the β-lactam ring but others, including aminoglycosides and chloramphenicol, are modified by the addition of chemical groups.

An understanding of the molecular mechanisms of drug resistance potentially allows the design and production of compounds which are less amenable as substrates for the modifying enzymes. Modern strategies for drug design and discovery have been reviewed by Kunz (1992). It remains to be seen whether a new generation of pharmaceuticals will arise from the application of crystallography and computational analysis. At present there is good reason to continue screening natural products for antibiotic activity and to look for active analogues of existing compounds; approaches which were responsible for the discovery of the vast majority of drugs on the market today.

6.4.1 Resistance to β-lactam antibiotics

β-Lactam antibiotics disrupt the synthesis of cell wall peptidoglycan by acylation of the active site serine of the DD-transpeptidase. Plasmid-mediated resistance involves production of β-lactamases which catalyse the hydrolysis and inactivation of the antibiotic (Frère, 1995). The labile four-member β-lactam ring which is a common feature of β-lactam antibiotic structure is shown in Fig. 6.6a. Hydrolysis breaks the ring, forming a stable penicilloic acid in the case of penicillin (Fig. 6.6b), or less stable products which undergo rapid breakdown to small fragments. The enzymes are widespread in the microbial world among both Gram-positive and Gram-negative organisms. The distribution of *bla* genes between chromosome and plasmid genomes is shown in Fig. 6.7. The majority of β-lactamases produced by Gram-positive organisms are extracellular, while those produced by Gram-negative organisms are, with few exceptions, periplasmic (i.e. retained between the inner and outer cell membranes). Because breakdown of the antibiotic takes place outside the cell, β-lactamase producing organisms can protect sensitive bacteria living in close proximity, especially if concentration of the drug is low.

6.4.2 Resistance to aminoglycosides

Aminoglycosides constitute a large group of agents which are of considerable importance in the treatment of serious infections. They include streptomycin, amikacin and kanamycin and exert their effects through interaction with ribosomes. A common structural

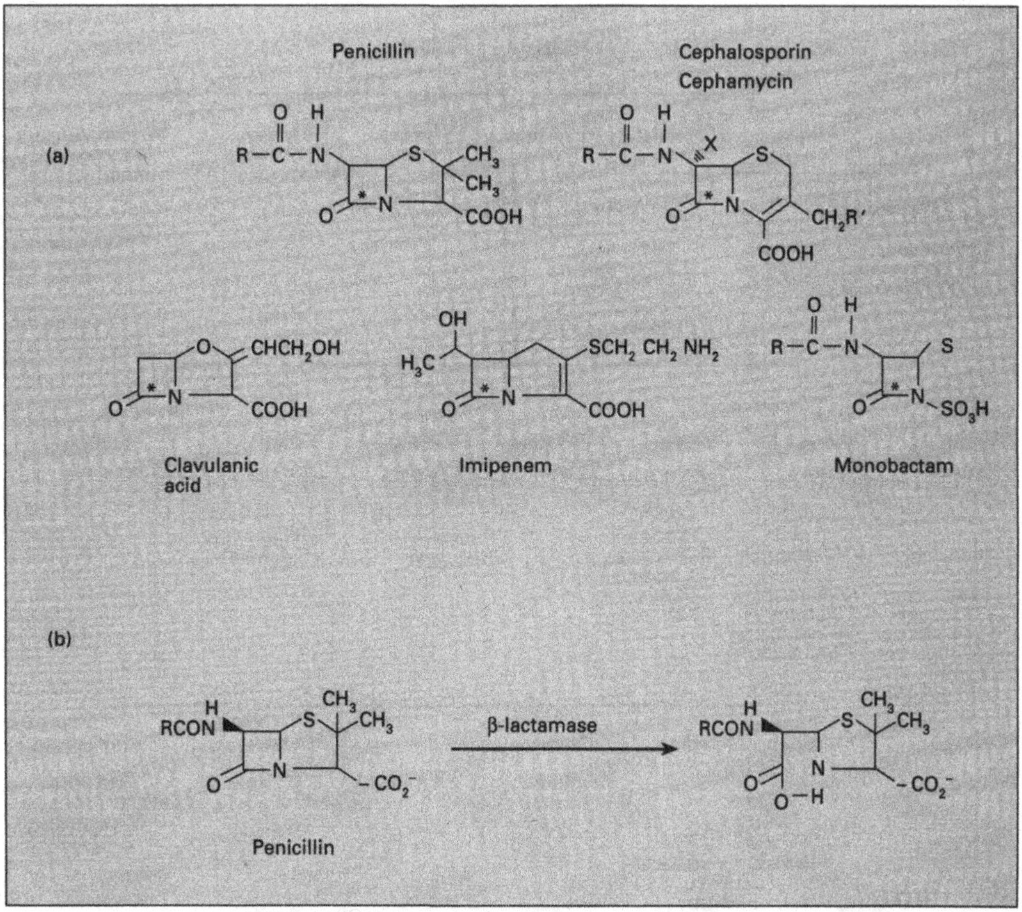

Fig. 6.6 β-Lactam antibiotics. (a) These molecules have in common an unstable four-membered ring of carbon and nitrogen atoms (indicated by an asterisk). (b) β-Lactamase catalyses the hydrolysis of the β-lactam ring, destroying the biological activity of the molecule. From Neu (1992).

feature is a cyclic alcohol in glycosidic linkage with amino-substituted sugars (Fig. 6.8). Plasmid-mediated resistance to the agents is widespread and almost always results from the production of enzymes which modify the antibiotic. The resistance genes often form part of a transposable element. Aminoglycoside modifying agents are capable of three basic reactions: N-acetylation, O-adenylylation and O-phosphorylation (Fig. 6.8). Within each group enzymes differ by the specific target of their attack and by the spectrum of substrates which are modified.

6.4.3 Resistance to the tetracyclines

The tetracyclines are a family of antibiotics introduced in the late 1940s which inhibit the binding of aminoacyl-tRNA to the A site on

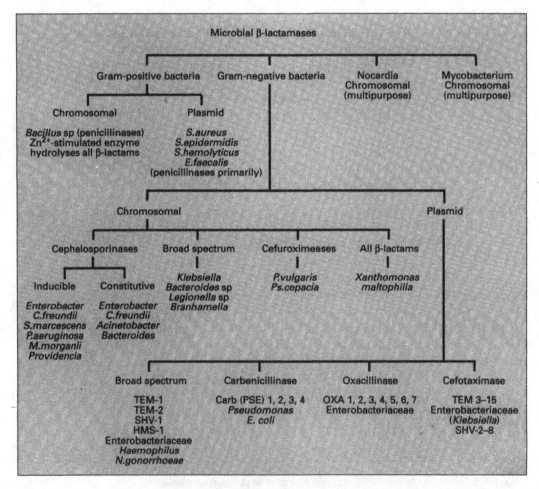

Fig. 6.7 The distribution of β-lactamases between chromosome and plasmid genomes. From Neu (1992).

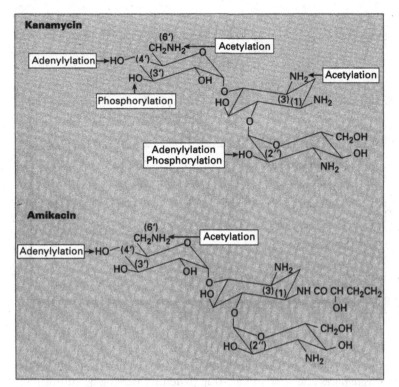

Fig. 6.8 Structures of kanamycin and amikacin. Arrows indicate the locations and chemical consequences of attack by modifying enzymes. From: *Plasmids* by Broda. Copyright © 1979 by W.H. Freeman and company. Used with permission.

the 30S ribosome. Broad-spectrum activity combined with high stability and relatively low toxicity led to their widespread use in clinical and veterinary medicine but their effectiveness has been seriously curtailed by the emergence of resistant organisms.

In 1983 the total annual production of tetracycline in the USA was 3.86 million kilograms, constituting 22% of total antibiotic production. However, only 29% of this was destined for use in disease treatment; the remainder was added to animal feed to stimulate weight gain. This practice, which was condemned by the Swann Report in 1969, has long been banned in Western Europe and most other countries of the world. The consequences of feedstock use are not surprising. By 1985, 12% of *E. coli* strains isolated from healthy human subjects in Boston were found to be resistant to tetracycline, and resistance among isolates from farm animals was as high as 80–90%. This cavalier attitude to antibiotic use does not only threaten the utility of tetracycline. The ubiquity of multiple drug resistance plasmids and the chemical stability of tetracyclines in feed and faeces means that their indiscriminate use selects for bacteria which are resistant to a range of structurally unrelated drugs.

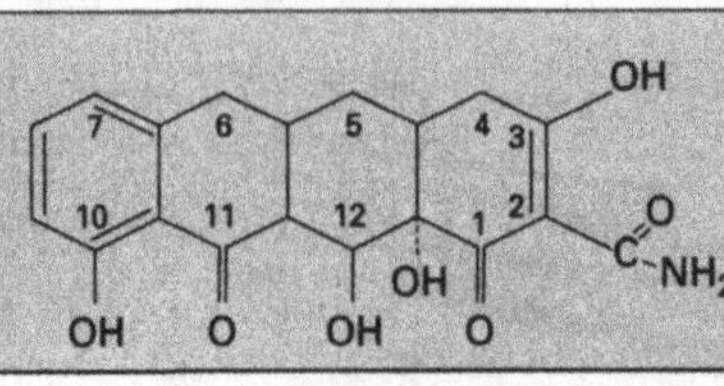

	Carbon atom substitutions		
Compound	C5	C6	C7
Chlortetracycline	H	CH$_3$OH	Cl
Oxytetracycline	OH	CH$_3$OH	H
Tetracycline	H	CH$_3$OH	H
Demeclocycline	H	OH	Cl
Methacycline	OH	CH$_2$	H
Doxycycline	OH	CH$_3$	H
Minocycline	H	H	N(CH$_3$)$_2$

Fig. 6.9 Structures of clinically important tetracyclines. Substitutions on carbon atoms 5, 6 and 7 are listed.

Tetracyclines are produced naturally by members of the genus *Streptomyces* and share a four-ring structure, varying by substitution at carbon atoms 5, 6 and 7 (Fig. 6.9). Tetracycline resistance determinants are widespread on plasmids, transposons and chromosomes in a very wide variety of species. Resistance genes in clinical strains are most often on plasmids or transposons and the variety of resistance determinants which have been discovered suggests that resistance has been evolving over millions of years as a response tetracycline production in nature.

Three major classes of tetracycline resistance determinants have been recognized on the basis of DNA sequence homology, the mechanism of resistance and the level of resistance (reviewed by Levy, 1984; Johnson and Adams, 1992). One class, found in Gram-negative organisms and including the resistance determinant of Tn*10* is inducible and involves the synthesis of a membrane-spanning TET protein which mediates efflux of the drug. A second class employing a different active efflux system is found in Gram-positive aerobic organisms including *Bacillus* and *Staphylococcus*. The third class (*tetM*), found in a diversity of organisms including mycoplasmas, the Gram-positive streptococci and Gram-negative species such as *Neisseria*, confers resistance by protecting ribosomes from inhibition by the drug. Among the streptococci, conjugative transposons such as Tn*916* mediate transfer of *tetM* in the absence of a conjugative plasmid and this may help to account for its widespread occurrence.

6.5 Plasmids and bacterial virulence

6.5.1 Virulence as a local adaptation

Antibiotic resistance is a conspicuous example of local adaptation which, consistent with the arguments of Eberhard described in chapter 2, is associated more often with plasmids than with the chromosome. Virulence in facultative pathogens is another example of a local or transient adaptation, since it is useful only as long as the bacterium persists with its pathogenic life style. In virulent *E. coli* which cause diarrhoea in man and animals, the host-specific fimbrial antigens which enable the bacterium to adhere to the intestinal epithelium are plasmid-encoded (Willshaw *et al.*, 1985). The relative ease with which plasmid-encoded genes can be gained and lost offers a pathogen the potential to fine-tune itself to a wide range of hosts; a degree of genetic flexibility which would not be available if the relevant genes were located on the chromosome.

What about obligate pathogens? Since the functions involved in the establishment and maintenance of infections are essential, the relevant genes might be expected to remain securely on the chromosome. This is partly true but many virulence genes assist with adaptation to the *local* environment rather than solving the more fundamental problems encountered by virulent organisms. Many obligate pathogens have more than one potential host and a set of alternative determinants is required to fine-tune each host–pathogen pair. The need to reshuffle these genes if a different host organism is infected means that they are likely to be plasmid-encoded. This is illustrated by the thousands of insecticidal strains of *B. thuringiensis*. No single isolate is active against all host species but different plasmid-encoded protoxins are effective against different groups of hosts (Aronson *et al.*, 1986).

6.5.2 The nature of virulence determinants

Only parasites which are poorly adapted to their host promote the violent reactions which are the symptoms of disease. The normal, long-term consequence of parasitism is the establishment of a stable relationship which places minimal demands upon the host, or even develops into symbiosis. This relatively happy state which exists for bacteria such as *E. coli* in the large intestine can be interrupted by the acquisition of plasmid-encoded virulence factors which provide a short-term advantage over competitors in the same niche (Brubaker, 1985). Plasmid-encoded virulence determinants have been implicated in adhesion, toxinogenesis, serum resistance, high-affinity iron transport, host-cell penetration and invasiveness. This is an area in which research is active and extremely diverse. The aim here is to illustrate the importance of plasmids in virulent organisms rather than to review the field in detail, so we will confine our attention (somewhat arbitrarily) to the virulence plasmids of *Salmonella typhimurium* and enterotoxigenic *E. coli*.

6.5.3 Virulence in *E. coli*

Escherichia coli is restricted normally to the lower part of the small intestine and to the large intestine. When pathogenic *E. coli* succeed in colonizing the small intestine the symptoms are similar to those of cholera, with a massive efflux of fluid and electrolytes from the secretory cells of the small intestine. Two broad groups of virulent *E. coli* (enterotoxigenic and enteropathogenic) are recognized and have been the subjects of extensive study.

Enteropathogenic *E. coli* (EPEC) cause enteric disease without toxin production. They attach to the intestinal mucosa and destroy the brush border microvilli, creating for themselves a pedestal of plasma membrane. EPEC affects mainly infants and is responsible for severe outbreaks of infantile gastroenteritis.

6.5.4 Enterotoxigenic *E. coli*

Enterotoxigenic *E. coli* (ETEC) pathogenicity is associated with bacterial invasion of the small intestine and enterotoxin production. Crucial determinants of both invasiveness and toxin production are plasmid-encoded. Enterotoxin proteins have a deleterious effect on the gastrointestinal tract; the most obvious clinical manifestations are diarrhoea and vomiting. ETEC are a major cause of diarrhoeal disease among travellers, in animals, and among infants in developing countries. The latter case is of particular concern because it is associated with alarmingly high mortality.

Various accessory genetic elements (transposons such as Tn*1681*, phage and plasmids) have been implicated in the production of *E. coli* and staphylococcal enterotoxins, Shiga-like toxins and cholera toxin (Betley *et al.*, 1986). Enterotoxigenic *E. coli* strains carry Ent plasmids encoding toxins which can be classified as either heat-labile (LT; inactivated after 30 min at 60°C) or heat-stable (ST; not inactivated after 30 min at 100°C). The plasmids form a diverse set, varying considerably in size and incompatibility grouping; some contain multiple origins. There is also variation in their capacity to encode additional functions including adhesin production, antibiotic resistance and conjugative transfer (Betley *et al.*, 1986; Neill and Holmes, 1988). Enterotoxigenic *E. coli* are responsible for mild to severe diarrhoea in humans and animals of all ages but toxin production alone is insufficient to cause disease symptoms if the bacteria are unable to invade the upper parts of the small intestine. Virulent enterotoxigenic strains must produce a colonization antigen (associated usually with fimbriae or pili on the bacterial surface), which facilitates invasion by enabling the bacterium to adhere to the small intestine.

Although they share many chemical and biochemical properties, colonization antigens are host-specific and provide a further example of plasmid encoded determinants fine-tuning the host–pathogen relationship. The adherence antigen associated with porcine enteropathogenic strains has been designated K88 while strains of bovine or ovine origin often carry the K99 antigen. Isolates of enteropathogenic *E. coli* from human sources express the CFAI and CFAII antigens. Because of their crucial role in virulence and potential uses as vaccines, the surface antigens of human pathogenic strains of *E. coli* have been studied extensively.

The genetic complexity of surface antigen production is illustrated by CFIIA which, although believed originally to be a single entity, is composed of three distinct antigens. CS1 and CS2 resemble pili while CS3 is finer and more flexible. Expression of all three antigens is dependent on the presence of a plasmid but only the genes encoding CS1 and CS3 are plasmid-located. Curiously, although the CS2 gene is located on the chromosome, its expression requires a plasmid-encoded positive regulator, Rns (Caron *et al.*, 1989), which also has a role in the expression of CS1.

6.5.5 *Salmonella* virulence plasmids

Members of the genus *Salmonella* are a well-known cause of gastroenteritis but are also responsible for systemic infections in humans and animals. Many virulence determinants are encoded on the *Salmonella* chromosome, including those required for adherence and invasion of mammalian epithelial cells, and resistance to phagocytic defences and complement-mediated bacteriolysis. Plasmid-free strains cause enteric infection and damage but are incapable of systemic infection. The presence of appropriate virulence plasmids enables salmonellae to increase their repertoire and establish progressive systemic infection of the reticuloendothelial organs in experimental animals. The *spv* (salmonella plasmid virulence) genes responsible for systemic infection and common to all members of the family of *Salmonella* virulence plasmids have been the subject of extensive molecular and genetic analysis (reviewed by Gulig *et al.*, 1993). The precise role of the *spv* gene products is unclear but cured bacteria appear less capable of proliferation in macrophages, which are believed to act as hosts for the growth and proliferation of bacteria in the reticuloendothelial system (Gulig and Doyle, 1993).

Salmonella virulence plasmids range in size from 50 to 90 kb, but a conserved region of 8 kb containing five genes (*spvRABCD*) is sufficient to restore the full spectrum of virulence to plasmid-cured strains. The genes constitute an operon which is transcriptionally silent in actively growing cells but is induced by the positive regulator SpvR in stationary phase and by extremes of pH and salt balance (Krause *et al.*, 1992; Valone *et al.*, 1993). This growth phase dependent pattern of control may represent the response of an intracellular pathogen to the host's attempt to starve the invader and provides another example of a plasmid underpinning adaptation to a local or transient environment.

Control of the *spv* genes involves a complicated regulatory network. Growth phase dependence is the consequence of a requirement for an alternative sigma factor (KatF or RpoS), which has been implicated in global changes to gene expression on entry into stationary phase. A *katF* mutant of *S. typhimurium* shows reduced expression of *spvABCD* and is attenuated for virulence (Fang *et al.*, 1992). The RpoS/KatF sigma factor is required for expression of the *spvR* gene product which positively regulates its own transcription. This unusual positive autoregulation is repressed by the products of *spvA* and *spvB* (Abe *et al.*, 1994).

References

Abe, A., Matsui, H., Danbara, H., Tanaka, K., Takahashi, H. & Kawahara, K. (1994). Regulation of *spvR* gene expression of *Salmonella* virulence plasmid pKDSC50 in *Salmonella choleraesuis* serovar Choleraesuis. *Molec. Microbiol.*, **12**, 779–787.

Abeles, A. L. (1986). P1 plasmid replication: purification and DNA-binding activity of the replication protein RepA. *J. Biol. Chem.*, **261**, 3548–3555.

Abeles, A. L. & Austin, S. J. (1991). Antiparallel plasmid–plasmid pairing may control P1 plasmid replication. *Proc. Natl. Acad. Sci. USA*, **88**, 9011–9015.

Abeles, A. L., Friedman, S. A. & Austin, S. J. (1985). Partition of unit-copy miniplasmids to daughter cells. III. The DNA sequence and functional organization of the P1 partition region. *J. Molec. Biol.*, **85**, 261–272.

Abeles, A. L., Reaves, L. D. & Austin, S. J. (1990). A single DnaA box is sufficient for initiation from the P1 plasmid origin. *J. Bacteriol.*, **172**, 4386–4391.

Abraham, E. P., Chain, E., Fletcher, C. M, Florey, H. W., Gardener, A. D., Heatley, N. G. & Jennings, M. A. (1941). Further observations on penicillin. *Lancet*, **ii**, 177–189.

Adams, D. E., Buska, J. B. & Cozzarelli, N. R. (1992). Cre-*lox* recombination in *Escherichia coli* cells. *J. Mol. Biol.*, **226**, 661–673.

Amabile-Cuevas, C. F. & Chicurel, M. E. (1992) Bacterial plasmids and gene flux *Cell*, **70**, 189–199.

Anderson, E. S. & Datta, N. (1965). Resistance to penicillins and its transfer in enterobacteriaceae. *Lancet*, **i**, 407–409.

Anderson, J. D., Gillespie, W. A. & Richmond, M. H. (1973) Chemotherapy and antibiotic-resistance transfer between enterobacteria in the human gastro-intestinal tract *J. Med. Microbiol.*, **6**, 461–473.

Argos, P., Landy, A., Abremski, K., Egan, J. B., Haggard-Ljungquist, E., Hoess, R. H., Kahn, M. L., Pierson, P., Sternberg, N. & Leong, J. M. (1986). The integrase family of site-specific recombinases: Regional similarities and global diversity *EMBO J.*, **5**, 433–440.

Aronson, A. I., Beckman, W. & Dunn, P. (1986). *Bacillus thuringiensis* and related insect pathogens. *Microbiol. Rev.*, **50**, 1–24.

Ataai, M. M. & Shuler, M. L. (1986). Mathematical model for the control of ColE1 type plasmid replication. *Plasmid*, **16**, 204–212.

Austin, S. J. (1988). Plasmid partition. *Plasmid*, **20**, 1–9.

Austin, S. & Abeles, A. (1983). Partition of unit-copy miniplasmids to daughter cells. II. The partition region of miniplasmid P1 encodes an essential protein and a centromere-like site at which it acts. *J. Molec. Biol.*, **169**, 373–387.

Austin, S. J. & Eichorn, B. G. (1992). Random diffusion can account for *topA*-dependent suppression of partition defects in low-copy-number plasmids. *J. Bacteriol.*, **174**, 5190–5195.

Austin, S. J. & Nordström, K. (1990). Partition-mediated incompatibility of bacterial plasmids. *Cell*, **60**, 351–354.

Austin S., Ziese M. & Sternberg, N. (1981). A novel role for site-specific recombination in maintenance of bacteria replicons. *Cell*, **25**, 729–736.

Austin, S., Friedman, S. & Ludtke, D. (1986). Partition functions of unit-copy plasmids can stabilize the maintenance of plasmid pBR322 at low copy number. *J. Bacteriol.*, **168**, 1010–1013.

References

Baquero, F., Lopez-Brea, M., Valls, A. & Canedo, T. (1977). *Chemotherapy (Basel)*, **23** (Suppl. 1), 133–140.

Barbour, A. G. & Garon, C. F. (1987a). Linear plasmids of the bacterium *Borrelia burgdorferi* have covalently closed ends. *Science*, **237**, 409–411.

Barbour, A. G. & Garon, C. F. (1987b). The genes encoding major surface proteins of *Borrelia burgdorferi* are located on a plasmid. *Ann. N. Y. Acad. Sci.*, **539**, 144–153.

Bates, E. E. & Gilbert, H. J. (1989). Characterization of a cryptic plasmid from *Lactobacillus plantanum*. *Gene*, **85**, 253–258.

Bauer, W. & Vinograd, J. (1968). The intercalation of closed circular DNA with intercalative dyes. *J. Molec. Biol.*, **33**, 141–171.

Bazaral, M. & Helinski, D. R. (1970). Replication of a bacterial plasmid and an episome in *Escherichia coli*. *Biochemistry*, **9**, 399–406.

Bech, F. W., Jorgensen, S. T., Diderichsen, B. & Karlstrom, O. H. (1985). Sequence of the *relB* transcription unit from *Escherichia coli* and identification of the *relB* gene. *EMBO J.*, **4**, 1059–1066.

Bedbrook, J. R. & Ausubel, F. M. (1976). Recombination between bacterial plasmids leading to the formation of plasmid multimers. *Cell*, **9**, 707–716.

Bennett, P. M., Grinsted, J., Choi, C. L. & Richmond, M. H. (1978). Characterization of Tn*501*, a transposon determining resistance to mercuric ions. *Molec. Gen. Genet.*, **159**, 101–106.

Benson, S. & Shapiro, J. (1978). TOL is a broad-host-range plasmid. *J. Bacteriol.*, **135**, 278–280.

Berg, C. M., Liu, L., Coon, M., Strausbaugh, L. D., Gray, P., Vartak, N. B., Brown, M., Talbot, D. & Berg, D. E. (1989). pBR322-derived multicopy plasmids harboring large inserts are often dimers in *Escherichia coli* K-12. *Plasmid*, **21**, 138–141.

Berg, D. E. (1990). Genomic rearrangements in prokaryotes. In B. D. H. & D. M. Glover (eds), *Gene Rearrangement* (pp. 1–50). Oxford: IRL Press.

Bergemann, A. D., Whitley, J. C. & Finch, L. R. (1989). Homology of mycoplasma plasmid pADB201 and staphylococcal plasmid pE194. *J. Bacteriol.*, **171**, 593–595.

Bergquist, P. L. (1987). *Incompatibility*. In *Plasmids a Practical Approach* (ed. K. G. Hardy), pp 37–78. Oxford: IRL Press.

Bergquist, P. L., Saadi, S. & Maas, W. K. (1986). Distribution of basic replicons having homology with RepFIA, RepFIB and RepFIC among IncF group plasmids. *Plasmid*, **15**, 19–34.

Bernander, R., Dasgupta, S. & Nordström, K. (1991). The *E. coli* cell cycle and the plasmid R1 replication cycle in the absence of the DnaA protein. *Cell*, **64**, 1145–1153.

Bernard, P. & Couturier, M. (1992). Cell killing by the F plasmid CcdB protein involves poisoning of DNA–topoisomerase II complexes. *J. Molec. Biol.*, **226**, 735–745.

Betley, M. J., Miller, V. L. & Mekalanos, J. J. (1986). Genetics of bacterial enterotoxins. *Ann. Rev. Microbiol.*, **40**, 577–605.

Bex, F., Karoui, H., Rokeach, L., Dreze, P., Garcia, L. & Couturier, M. (1983). Mini-F encoded proteins: identification of a new 10.5 kilodalton species. *EMBO J.*, **2**, 1853–1861.

Bickle, T. A. & Krüger, D. H. (1993). Biology of DNA restriction. *Microbiol. Rev.*, **57**, 434–450.

Birnboim, H. C. & Doly, J. (1979). A rapid alkaline extraction procedure screening recombinant plasmid DNA. *Nucleic Acids Res.*, **7**, 1513–1523.

Bissonnette, L. & Roy, P. H. (1992). Characterization of InO of *Pseudomonas aeruginosa* pVS1, an ancestor of integrons of multiresistance plasmids and transposons of Gram-negative bacteria. *J. Bacteriol.*, **174**, 1248–1257.

Biswas, G. D., Burnstein, K. & Sparling, P. F. (1985). Plasmid Transformation in Neisseria gonorrhoeae. In *The Pathogenic Neisseriae* (ed. G. K. Schoolnile), pp. 204–208. Washington DC: American Society for Microbiology.

Blakely, G., Colloms, S., May, G., Burke, M. & Sherratt, D. (1991). *Escherichia coli* XerC recombinase is required for chromosomal segregation at cell division. *New Biologist*, 3, 789–798.

Blakely, G. W., May, G., McCulloch, R., Arciszewska, L. K., Burke, M., Lovett, S. T. & Sherratt, D. J. (1993). Two related recombinases are required for site-specific recombination at *dif* and *cer* in *Escherichia coli* K12. *Cell*, 75, 351–361.

Blenkharn, J. I. & Hughes, V. M. (1982). Suction apparatus and hospital infection due to multiply-resistant *Klebsiella aerogenes*. *J. Hosp. Infect.*, 3, 173–178.

Blomberg, P., Wagner, E. G. H. & Nordström, K. (1990). Control of replication of plasmid R1: the duplex between the antisense RNA, CopA, and its target, CopT, is processed specifically *in vivo* and *in vitro* by RNase III. *EMBO J.*, 9, 2331–2340.

Blomberg, P., Nordström, K. & Wagner, E. G. H. (1992). Replication control of plasmid R1: RepA synthesis is regulated by CopA RNA through inhibition of leader peptide translation. *EMBO Journal*, 11, 2675–2683.

Blomberg, P., Engdahl, H. M., Malmgren, C., Romby, P. & Wagner, E. G. H. (1994). Replication control of plasmid R1: disruption of an inhibitory RNA structure that sequesters the *repA* ribosome-binding site permits *tap*-independent RepA synthesis. *Molec. Microbiol.*, 12, 49–60.

Boe, L., Gerdes, K. & Molin, S. (1987). Effects of genes exerting growth inhibition and plasmid stability on plasmid maintenance. *J. Bacteriol.*, 169, 4646–4650.

Boles, T. C., White, J. H. & Cozzarelli, N. R. (1990). Structure of plectonemically supercoiled DNA. *J. Molec. Biol.*, 213, 931–951.

Bolivar, F., Rodriguez, R. L., Green, P. J., Betlach, M. C., Heyneker, H. L., Boyer, H. W., Crosa, J. H. & Falkow, S. (1977). Construction and characterization of new cloning vehicles. II. A multipurpose cloning system. *Gene*, 2, 95–113.

Bouma, J. E. & Lenski, R. E. (1988). Evolution of a bacteria/plasmid association. *Nature*, 335, 351–352.

Bouthier de la Tour, C., Portemer, C., Nadal, M., Stetter, K. O., Forterre, P. & Duguet, M. (1990). Reverse gyrase, a hallmark of the hyperthermophilic Archaebacteria. *J. Bacteriol.*, 172, 6803–6808.

Boyd, A. C., Archer, J. A. C. & Sherratt, D. J. (1989). Characterization of the ColE1 mobilization region and its protein products. *Molec. Gen. Genet.*, 217, 488–498.

Bradley, D. E., Taylor, D. E. & Cohen, D. R. (1980). Specification of surface mating systems among conjugative drug resistance plasmids in *Escherichia coli* K-12. *J. Bacteriol.*, 143, 1466–1470.

Bramhill, D. & Kornberg, A. (1988a). A model for initiation at origins of DNA replication. *Cell*, 54, 915–918.

Bramhill, D. & Kornberg, A. (1988b). Duplex opening by DnaA protein at novel sequences in initiation of replication at the origin of the *E. coli* chromosome. *Cell*, 52, 743–755.

Bravo, A., de Torrontegui, G. & Diaz, R. (1987). Identification of components of a new stability system of plasmid R1, ParD, that is close to the origin of replication of this plasmid. *Molec. Gen. Genet.*, 210, 101–110.

Bravo, A., Ortega, S., de Torrontegui, G. & Diaz, R. (1988). Killing of *Escherichia coli* cells modulated by components of the stability system ParD of plasmid R1. *Molec. Gen. Genet.*, 215, 146–151.

Bremer, H., Churchward, G. & Young, R. (1979). Relation between growth and replication in bacteria. *J. Theor. Biol.*, 81, 533–545.

References

Brenner, M. & Tomizawa, J. (1991). Quantitation of ColE1-encoded replication elements. *Proc. Natl. Acad. Sci.*, **88**, 405–409.

Broda (1980) *Plasmids in Human and Veterinary Medicine* (Chapter 6) W. H. Freeman, Oxford.

Brolle, D. F., Pape, H., Hopwood, D. A. & Kieser, T. (1993). Analysis of the transfer region of the Streptomyces plasmid SCP2*. *Molec. Microbiol.*, **10**, 157–170.

Brooks Low, K. (1987). HFr strains of *Escherichia coli K-12*. In *Escherichia coli and Salmonella typhimurium cellular and molecular Biology* (ed. F. C. Neidhardt), pp. 1134–1137. Washington DC: American Society for Microbiology.

Brubaker, R. R. (1985). Mechanisms of bacterial virulence. *Ann. Rev. Microbiol.*, **39**, 21–50.

Buchanan-Wollaston, V., Passiatore, J. E. & Cannon, F. (1987). The *mob* and *oriT* mobilization functions of a bacterial plasmid promote its transfer to plants. *Nature*, **328**, 172–175.

Cannon, P. M. & Strike, P. (1992). Complete nucleotide sequence and gene organization of plasmid NTP16. *Plasmid*, **27**, 220–230.

Carle, G. R., Frank, M. & Olson, M. V. (1986). Electrophoretic separations of large DNA molecules by periodic inversion of the electric field. *Science*, **232**, 65–68.

Carleton, S., Projan, S. J., Highlander, S. K., Moghazeh, S. & Novick, R. P. (1984). Control of pT181 replication II. Mutational analysis. *EMBO J.*, **3**, 2407–2414.

Caron, J., Coffield, L. M. & Scott, J. R. (1989). A plasmid-encoded regulatory gene, rns, required for the expression of the CS1 and CS2 adhesins of enterotoxigenic *Escherichia coli*. *Proc. Natl. Acad. Sci. USA*, **86**, 963–967.

Cesareni, G., Meusing, M. A. & Polisky, B. (1982). Control of ColE1 DNA replication: the *rop* gene product negatively affects transcription from the replication primer promoter. *Proc. Natl. Acad. Sci. USA*, **79**, 6313–6317.

Cesareni, G., Helmer-Citterich, M. & Castagnoli, L. (1991). Control of ColE1 plasmid replication by antisense RNA. *TIG*, **7**, 230–235.

Chang, A. C. Y. & Cohen, S. N. (1978). Construction and characterization of amplifiable multicopy DNA cloning vehicles derived from the p15A cryptic miniplasmid. *J. Bacteriol.*, **134**, 1141–1156.

Chao, L. & Levin, B. R. (1981). Structured habitats and the evolution of anticompetitor toxins in bacteria. *Proc. Natl. Acad. Sci. USA*, **78**, 6324–6328.

Chattoraj, D. K., Snyder, K. M. & Abeles, A. L. (1985). P1 plasmid replication: multiple functions of RepA protein at the origin. *Proc. Natl. Acad. Sci. USA*, **82**, 2588–2592.

Chattoraj, D. K., Mason, R. J. & Wickner, S. H. (1988). Mini-P1 plasmid replication: the autoregulation–sequestration paradox. *Cell*, **52**, 551–557

Chau, P. Y., Ling, J., Threlfall, E. J. & Im, S. W. K. (1982). Genetic instability of R plasmids in relation to the shift of drug resistance patterns in *Salmonella johannesburg J. Gen. Microbiol.*, **128**, 239–245.

Chea, U. E., Weigand, W. A. & Stark, B. C. (1987). Effects of recombinant plasmid size on cellular processes in *Escherichia coli*. *Plasmid*, **18**, 127–134.

Chiang, C. S. & Bremer, H (1988) Stability of pBR322-derived plasmids *Plasmid*, **20**, 207–220.

Chiang, C. & Bremer, H. (1991). Maintenance of pBR322-derived plasmids without functional RNAI. *Plasmid*, **26**, 186–200.

Clark, A. J. (1973). Recombination-deficient mutations of *E. coli* and other bacteria. *Ann. Rev. Genet.*, **7**, 67–86

Clark, A. J. (1980). A view of the RecBC and RecF pathways of *E. coli* recombination. In B. Alberts & C. F. Fox (eds), *Mechanistic Studies of DNA Replication and Genetic Recombination* (pp. 891–899). New York: Academic Press.

Clewell, D. B., Flanagan, S. E., Zitow, L. A., Su, Y. A., He, P., Senghas, E. & Weaver, K. (1991). Properties of Conjugative Transposon Tn916. In *Genetics and Molecular Biology of Streptococci, Lactococci, and Enterococci* (eds. G. Dunny, P. P. Cleary & L. L. Mckay), pp. 39–44. Washington DC: American Society for Microbiology.

Clewell, D. B. (1993). Bacterial sex pheromone-induced plasmid transfer. *Cell*, **73**, 9–12.

Clewell, D. B. & Flannagan, S. E. (1993). The conjugative transposons of Gram-positive bacteria. In *Bacterial Conjugation* (ed. D. B. Clewell), pp. 369–393. New York: Plenum Press.

Cohen, A. & Laban, A. (1983). Plasmidic recombination in *Escherichia coli* K12: the role of *recF* gene function. *Molec. Gen. Genet.*, **189**, 471–474.

Cohen, M. L. (1992). Epidemiology of drug resistance: implications for a post-antimicrobial era. *Science*, **257**, 1050–1055.

Collis, C. M. & Hall, R. M. (1992a). Site-specific deletion and rearrangement of integron insert genes catalysed by the integron DNA integrase. *J. Bacteriol.*, **174**, 1574–1585.

Collis, C. M. & Hall, R. M. (1992b). Gene cassettes from the insert region of integrons are excised as covalently closed circles. *Molec. Microbiol.*, **6**, 2875–2885.

Collis, C. M., Grammaticopoulos, G., Briton, J., Stokes, H. W. & Hall, R. M. (1993). Site-specific insertion of gene cassettes into integrons. *Molec. Microbiol.*, **9**, 41–52.

Colloms, S. D., Sykora, P., Szatmari, G. & Sherratt, D. J. (1990). Recombination at ColE1 *cer* requires the *Escherichia coli xerC* gene product, a member of the lambda integrase family. *J. Bacteriol*, **172**, 6973–6980.

Conley, D. L. & Cohen, S. N. (1995). Effects of the pSC101 partition *(par)* locus on *in vivo* DNA supersoiling near the plasmid replication origin. *Nucl. Acids Res.*, **23**, 701–707.

Costerton, J. W., Cheng, K. J., Gesey, G. G., Ladd, T. I., Nickel, J. C., Dasgupta, M. & Marrie, T. J. (1987). Bacterial biofilms in nature and disease. *Ann. Rev. Microbiol.*, **41**, 435–464.

Couturier, M., Bex, F., Bergquist, P. L. & Maas, W. K. (1988). Identification and classification of bacterial plasmids. *Microbiol. Rev.*, **52**, 375–395.

Dasgupta, S., Masukata, H. & Tomizawa, J. (1987). Transcriptional activation of ColE1 DNA synthesis by displacement of the non-transcribed strand. *Cell*, **51**, 1123–1130.

Datta, N. (1984). Plasmids of Enteric Bacteria. In *Antimicrobial Drug Resistance* (ed. L. E. Bryan), pp. 487–496. Orlando: Academic Press.

Datta, N. (1985). Plasmids as Organisms. In *Plasmids in Bacteria* (eds. D. R. Helinski, S. N. Cohen, B. Clewell, D. A. Jackson & A. Hollaender) pp. 3–16. New York: Plenum Press.

Davey, R. P. & Reanney, D. C. (1980). Extrachromosomal Genetic Elements and the Adaptive Evolution of Bacteria. In *Evolutionary Biology* (eds. M. K. Hecht, W. C. Steere & B. Wallace), pp. 113–147. New York: Plenum Press.

Davis, M. A. & Austin, S. J. (1988). Recognition of the P1 plasmid centromere analog involves binding of the ParB protein and is modified by a specific host factor. *EMBO J.*, **7**, 1881–1888.

Davis, M. A., Martin, K. A. & Austin, S. J. (1990). Specificity switching of the P1 plasmid centromere-like site. *EMBO J.*, **9**, 991–998.

Davis, M. A., Martin, K. A. & Austin, S. J. (1992). Biochemical activities of the ParA protein of the P1 plasmid. *Molec. Microbiol.*, **6**, 1141–1147.

Day, M. & Fry, J. C. (1992). Gene Transfer in the Environment: Conjugation. In *Release of genetically engineered and other micro-organisms* (eds. J. C. Fry & M. J. Day) pp. 40–53. Cambridge, UK: Cambridge University Press.

de Feyter, R., Wallace, C. & Lane, D. (1989). Autoregulation of the *ccd* operon in the F plasmid. *Molec. Gen. Genet.*, **218**, 481–486.

References

de la Cruz, F. & Grinsted, J. (1982). Genetic and molecular characterization of Tn21, a multiple resistance transposon from R100–1. *J. Bacteriol.*, **151**, 222–228.

del Solar, G., Moscoso, M. & Espinosa, M. (1993). Rolling circle-replicating plasmids from Gram-positive and Gram-negative bacteria: a wall falls. *Molec. Microbiol.*, **8**, 789–796.

Delver, E. P., Kotova, V. U., Zavilgelsky, G. B. & Belogurov, A. A. (1991). Nucleotide sequence of the gene (*ard*) encoding the antirestriction protein of plasmid ColIb-P9. *J. Bacteriol.*, **173**, 5887–5892.

Demerec, M. (1948). Origin of bacterial resistance to antibiotics. *J. Bacteriol.*, **56**, 63–74.

Derbyshire, K. M., Hatfull, G. & Willetts, N. (1987). Mobilization of the non-conjugative plasmid RSF1010: genetic and DNA sequence analysis of the mobilization region. *Molec. Gen. Genet.*, **206**, 161–168.

Di Laurenzio, L., Frost, L. S. & Paranchych, W. (1992). The TraM protein of the conjugative plasmid F binds to the origin of transfer of the F and ColE1 plasmids. *Molec. Microbiol.*, **6**, 2951–2959.

Dodd, H. & Bennett, P. M. (1986). Location of the site-specific recombination of R46: a function necessary for plasmid maintenance. *J. Gen. Microbiol.*, **132**, 1009–1020.

Dotto, G. P., Horiuchi, K. & Zinder, N. D. (1984). The functional origin of bacteriophage f1 DNA replication. Its signals and domains. *J. Molec. Biol.*, **172**, 507–521.

Durland R. H. & Helinski, D. R. (1990). Replication of the broad-host-range plasmid RK2: direct measurement of intracellular concentrations of the essential TrfA replication proteins and their effect on plasmid copy number. *J. Bacteriol.*, **172**, 3849–3858.

Eberhard, W. G. (1990). Evolution in bacterial plasmids and levels of selection. *Quart. Rev. Biol.*, **65**, 3–22.

Eguchi, Y., Itoh, T. & Tomizawa, J. (1991). Antisense RNA. *Ann. Rev. Biochem.*, **60**, 631–652.

Ehrlich, P. (1907). Chemotherapeutische Trypanosomen-Studien. *Ber. Klin. Wochschr.*, **44**, 233–236.

Eraso, J. M. & Weinstock, G. M. (1992). Anaerobic control of colicin E1 production. *J. Bacteriol.*, **174**, 5101–5109.

Ezaki, B., Ogura, T., Niki, H. & Hiraga, S. (1991). Partitioning of a mini-F plasmid into anucleate cells of the *mukB* null mutant, *J. Bacteriol.*, **173**, 6643–6646.

Fang, F. C., Libby, S. J., Buchmeier, N. A., Loewen, P. C., Switala, J., Harwood, J., & Guiney, D. G. (1992). The alternative σ factor KatF (RpoS) regulates *Salmonella* virulence. *Proc. Natl. Acad. Sci. USA*, **89**, 11978–11982.

Ferdows, M. S. & Barbour, A. G. (1989). Megabase-sized linear DNA in the bacterium *Borrelia burgdorferi*, the Lyme disease agent. *Proc. Natl. Acad. Sci. USA*, **86**, 5969–5973.

Figurski, D. H., Meyer, R. J. & Helinski, D. R. (1979). Suppression of ColE1 replication properties by the IncP-1 plasmid RK2 in hybrid plasmids constructed *in vitro. J. Mol. Biol.* **133**, 295–318.

Firth, N. & Skurray, R. (1992). Characterization of the F plasmid bifunctional conjugation gene, *traG. Molec. Gen. Genet.*, **232**, 145–153.

Fishel, R. A., James, A. A. & Kolodner, R. (1981). recA-independent general genetic recombination of plasmids. *Nature*, **294**, 184–186.

Fitzwater, T., Yang, Y.-L., Zhang, X.-Y. & Polisky, B. (1992). Mutations affecting RNA–DNA hybrid formation of the ColE1 replication primer RNA. Restoration of the RNA I sensitivity to a copy-number mutant by second-site mutations. *J. Molec. Biol.*, **226**, 997–1008.

Flaks, J. G., Leboy, P. S., Birge, E. A. & Kurland, C. G. (1966). Mutations and genetics concerned with the ribosome. *Cold Spring Harbor Symp. Quant. Biol.*, **31**, 623–631.

Fleming, M. P., Datta, N. & Gruneberg, R. N. (1972). *Br. Med. J.*, **1**, 726–728.

Fox, J. L. (1992). *Am. Soc. Microbiol. News*, **58**, 135.

Franke, A. E. & Clewell, D. B. (1981). Evidence for a chromosome-borne resistance transposon (Tn916) in *Streptococcus faecalis* that is capable of conjugal transfer in the absence of a conjugative plasmid. *J. Bacteriol.*, **145**, 494–502.

Frère, J.-M. (1995). Beta-lactamases and bacterial resistance to antibiotics. *Molec. Microbiol.*, **16**, 385–395.

Freundlich, M., Ramani, N., Mathew, E., Sirko, A. & Tsui, P. (1992). The role of integration host factor in gene expresion in *Escherichia coli*. *Molec. Microbiol.*, **6**, 2557–2563.

Friedman, D. I. (1988). Integration host factor: a protein for all reasons. *Cell*, **55**, 545–554.

Friedman, S. A. & Austin, S. J. (1988). The P1 plasmid–partition system synthesizes two essential proteins from an autoregulated operon. *Plasmid*, **19**, 103–112.

Frost, L., Lee, S., Yanchar, N. & Paranchych, W. (1989). *finP* and *fisO* mutations in FinP anti-sense RNA suggest a model for FinOP action in the repression of bacterial conjugation by the Flac plasmid JCFL0. *Molec. Gen. Genet.*, **218**, 152–160.

Fuller, R. S., Funnell, B. E. & Kornberg, A. (1984). The DnaA protein complex with the *E. coli* chromosomal replication origin (*oriC*) and other DNA sites. *Cell*, **38**, 889–900.

Funnell, B. E. (1988a). Mini-P1 plasmid partitioning: excess ParB protein destabilizes plasmids containing the centromere *parS*. *J. Bacteriol.*, **170**, 954–960.

Funnell, B. E. (1988b). Participation of *Escherichia coli* integration host factor in the P1 plasmid partitioning system. *Proc. Natl. Acad. Sci. USA*, **85**, 6657–6661.

Funnell, B. E. & Gagnier, L. (1995). Partition of P1 plasmids in *Escherichia coli mukB* chromosomal partition mutants. *J. Bacteriol.*, **177**, 2381–2386.

Gelfand, D. H., Shepard, H. M., O'Farrell, P. H. & Polisky, B. (1978). Isolation and characterization of a ColE1-derived plasmid copy number mutant. *Proc. Natl. Acad. Sci. USA*, **75**, 5869–5873.

Gennaro, M. L. & Novick, R. P. (1988). An enhancer of DNA replication. *J. Bacteriol.*, **170**, 5709–5717.

Gerdes, K. & Molin, S. (1986). Partitioning of plasmid R1. Structural and functional analysis of the *parA* locus. *J. Molec. Biol.*, **190**, 269–279.

Gerdes, K., Rasmussen, P. B. & Molin, S. (1986a). Unique type of plasmid maintenance function: postsegregational killing of plasmid-free cells. *Proc. Natl. Acad. Sci. USA*, **83**, 3116–3120.

Gerdes, K., Bech, F. W., Jorgensen, S. T., Lobner-Olesen, A., Rasmussen, P. B., Atlung, T., Boe, L., Karlstrom, O., Molin, S. & von Meyenberg, K. (1986b). Mechanism of postsegregational killing by the *hok* gene product of the *parB* system of plasmid R1 and its homology with the *relF* gene product of the *E. coli relB* operon. *EMBO J.*, **5**, 2023–2029.

Gerdes, K., Helin, K., Christensen, O. W. & Lobner-Olesen, A. (1988). Translational control and differential RNA decay are key elements regulating postsegregational expression of the killer protein encoded by the *parB* locus of plasmid R1. *J. Molec. Biol.*, **203**, 119–129.

Gerdes, K., Poulson, L. K., Thistead, T., Nielsen, A. K., Martinussen, J. & Andreasen, P. H. (1990). The *hok* killer gene family in Gram-negative bacteria. *New Biologist*, **2**, 946–956.

Gerdes, K., Nielsen, A., Thorstead, P. & Wagner, E. G. H. (1992). Mechanism of killer gene activation. Antisense RNA-dependent RNase III cleavage ensures rapid turn-over of the stable Hok, SrnB and PndA effector messenger RNAs. *J. Molec. Biol.*, **226**, 637–649.

Gerlitz, M., Hrabak, O. & Schwab, H. (1990). Partitioning of broad-host-range plasmid RP4 is a complex system involving site-specific recombination. *J. Bacteriol.*, **172**, 6194–6203.

Gille, H. & Messer, W. (1991). Localized DNA melting and structural pertubations in the origin of replication, oriC, of *Escherichia coli in vitro* and *in vivo*. *EMBO J.*, **10**, 1579–1584.

Gillen, J. R., Willis, D. K. & Clark, A. J. (1981). Genetic analysis of the *recE* pathway of genetic recombination in *Escherichia coli* K-12. *J. Bacteriol.*, **145**, 521–532.

Golub, E. I. & Panzer, H. A. (1988). The F factor of *Escherichia coli* carries a locus of stable plasmid inheritance *stm*, similar to the *parB* locus of plasmid R1. *Molec. Gen. Genet.*, **214**, 353–357.

Gordon, D. M. (1992). Rate of plasmid transfer among *Escherichia coli* strains isolated from natural populations. *J. Gen. Microbiol.*, **138**, 17–21.

Green, P. J., Gupta M., Boger, H. W., Brown, W. E. & Rosenberg, J. M. (1981). Sequence analysis of the DNA encoding the EcoRI endonuclease and methylase. *J. Biol. Chem.*, **256**, 2143–2153.

Griffith, F. (1928). The significance of pneumococcal types. *J. Hyg.*, **27**, 113–159.

Grindey, N. D. F. & Reed, R. R. (1985). Transpositional recombination in prokaryotes. *Ann. Rev. Biochem.*, **54**, 863–896.

Grinsted, J. (1986). Evolution of transposable elements. *J. Antimicrob. Chemother.*, **18**, 77–83.

Grinsted, J. & Bennett, P. M. (1988). Preparation and Electrophoresis of Plasmid DNA. In *Methods in Microbiology* (eds. J. Grinsted & P. M. Bennett), pp. 129–142. London: Academic Press.

Grinsted, J., de la Cruz, F. & Schmitt, R. (1990). The Tn21 subgroup of bacterial transposable elements. *Plasmid*, **24**, 163–189.

Gruss, A. & Ehrlich, S. D. (1989). The family of highly interrelated single-stranded deoxyribonucleic acid plasmids. *Microbiol. Rev.*, **53**, 231–241.

Gruss, A., Ross, H. F. & Novick, R. P. (1987). Functional analysis of a palindromic sequence required for normal replication of several staphylococcal plasmids. *Proc. Natl. Acad. Sci. USA*, **84**, 2165–2169.

Guiney, D. G. & Lanka, E. (1989). Conjugative Transfer of IncP Plasmids. In *Promiscuous Plasmids of Gram-Negative Bacteria* (ed. C. M. Thomas), pp. 27–56. London: Academic Press.

Gulig, P. A., Danbara, H., Guiney, D. G., Lax, A. J., Norel, F. & Rhen, M. (1993). Molecular analysis of *spv* virulence genes of the salmonella virulence plasmids. *Molec. Microbiol.*, **7**, 825–830.

Gulig, P. A. & Doyle, T. J. (1993). The *Salmonella typhimurium* virulence plasmid increases the net growth rate of salmonellae in mice. *Infect. Immun.*, **61**, 504–511.

Hakkaart, M. J. J., van den Elzen, P. J. M., Veltkamp, E. & Nijkamp, H. J. J. (1984). Maintenance of multicopy plasmid CloDF13 in *E. coli* cells: evidence for site-specific recombination at *parB*. *Cell*, **36**, 203–209.

Hall, R. M. & Stokes, H. W. (1993). Integrons: novel DNA elements which capture genes by site-specific recombination. *Genetica*, **90**, 115–132.

Hall, R. M., Brookes, D. E. & Stokes, H. W. (1991). Site-specific insertion of genes into integrons: the role of the 59-base element and determination of the recombination cross-over point. *Molec. Microbiol.*, **5**, 1941–1959.

Hardy, K. G. (1987). Purification of Bacterial Plasmids. In *Plasmids, a Practical Approach* (ed. K. G. Hardy), pp. 1–6. Oxford: IRL Press.

Harrison, J. L., Taylor, I. M., Platt, K. & O'Connor, C. D. (1992). Surface exclusion specificity of the TraT lipoprotein is determined by single alterations in a five-amino-acid region of the protein. *Molec. Microbiol.*, **6**, 2825–2832.

Hayes, W. (1953). Observations on a transmissible agent determining sexual differentiation in *Bact. coli*. *J. Gen. Microbiol.*, **8**, 72–88.

He, L., Söderbom, F., Wagner, E. G. H., Binnie, U., Binns, N. & Masters, M. (1993). PcnB is required for the rapid degradation of RNAI, the antisense RNA that controls the copy number of ColE1-related plasmids. *Molec. Microbiol.*, **9**, 1131–1142.

Heinemann, J. A. & Ankenbauer, R. G. (1993). Retrotransfer of IncP plasmid R751 from *Escherichia coli* maxicells: evidence for the genetic sufficiency of self-transferable plasmids for bacterial conjugation. *Molec. Microbiol.*, **10**, 57–62.

Heinemann, J. A. & Sprague, G. F. (1989). Bacterial conjugative plasmids mobilize DNA transfer between bacteria and yeast. *Nature*, **340**, 205–209.

Helmer-Citterich, M., Anceschi, M. M., Banner, D. W. & Cesareni, G. (1988). Control of ColE1 replication: low affinity specific binding of Rop (Rom) to RNAI and RNAII. *EMBO J.*, **7**, 557–566.

Highlander, S. K. & Novick, R. P. (1987). Plasmid repopulation kinetics in *Staphylococcus aureus*. *Plasmid*, **17**, 210–221.

Highlander, S. K. & Novick, R. P. (1990). Mutational and physiological analyses of plasmid pT181 functions expressing incompatibility. *Plasmid*, **23**, 1–15.

Hildebrandt, E. R. & Cozzarelli, N. R. (1995). Comparison of recombination *in vitro* and in *E. coli* cells: measure of the effective concentration of DNA *in vivo*. *Cell*, **81**, 331–340.

Hinnebusch, J. & Barbour, A. G. (1991). Linear plasmids of *Borrelia burgdorferi* have a telomeric structure and sequence similar to those of a eukaryotic virus. *J. Bacteriol.*, **173**, 7233–7239.

Hinnebusch, J. & Barbour, A. G. (1992). Linear- and circular-plasmid copy numbers in *Borrelia burgdorferi*. *J. Bacteriol.*, **174**, 5251–5257.

Hinnesbusch, J. & Tilly, K. (1993). Linear plasmids and chromosomes in bacteria. *Molec. Microbiol.*, **10**, 917–922.

Hinnebusch, J., Bergstrom, S. & Barbour, A. G. (1990). Cloning and sequence analysis of linear plasmid telomeres of the bacterium *Borrelia burgdorferi*. *Molec. Microbiol.*, **4**, 811–820.

Hiraga, S., Jaffe, A., Ogura, T., Mori, H. & Takahashi, H. (1986). F plasmid ccd mechanism in *Escherichia coli*. *J. Bacteriol.*, **166**, 100–104.

Hirochika, H., Nakamura, K. & Sakaguchi, K. (1984). DNA plasmid from *Streptomyces rochei* with an inverted repetition of 614 base pairs. *EMBO J.*, **3**, 761–766.

Hirose, S., Hiraga, S. & Okazaki, T. (1983). Initiation site of deoxyribonucleotide polymerization at the replication origin of the *Escherichia coli* chromosome. *Molec. Gen. Genet.*, **189**, 422–431.

Holmes, D. S. & Quigley, M. (1981). A rapid boiling method for the preparation of bacterial plasmids. *Anal. Biochem.*, **114**, 193–197.

Hopwood, D. A. & Kieser, T. (1993). Conjugative Plasmids in Streptomyces. In *Bacterial Conjugation* (ed. D. B. Clewell), pp. 293–311. New York: Plenum Press.

Horii, Z. I. & Clark, A. J. (1973). Genetic analysis of the RecF pathway to genetic recombination in *Escherichia coli* K-12: isolation and characterization of mutants. *J. Molec. Biol.*, **80**, 327–344.

Hsu, J., Bramhill, D. & Thompson, C. M. (1994). Open complex formation by DnaA initiation protein at the *Escherichia coli* chromosomal origin requires the 13-mers precisely spaced relative to the 9-mers. *Molec. Microbiol.*, **11**, 903–911.

Hwang, D. S. & Kornberg, A. (1990). A novel protein binds a key origin sequence to block replication of an *E. coli* minichromosome. *Cell*, **63**, 325–331.

References

Inouye, M. & Inouye, S. (1992). Retrons and multicopy single-stranded DNA. *J. Bactoriol.*, **174**, 2419–2424.

Iordanescu, S. (1995). Plasmid pT181 replication is decreased at high levels of RepC per plasmid copy. *Molec. Microbiol.*, **16**, 477–484.

Ippen-lhler, K. A. & Minkley, E. G. (1986). The conjugation system of F, the fertility factor of *Escherichia coli*. *Ann. Rev. Genet.*, **20**, 593–624.

Ippen-lhler, K. (1989). Bacterial Conjugation. In *Gene Transfer in the Environment* (ed. S. B. Levy & R. V. Miller), pp. 32–72. New York: McGraw-Hill.

Itoh, T. & Tomizawa, J. (1978). Initiation of replication of plasmid ColE1 DNA by RNA polymerase, ribonuclease H and DNA polymerase 1. *Cold Spring Harbor Symp. Quant. Biol.*, **43**, 409–418.

Itoh, T. & Tomizawa, J. (1980). Formation of an RNA primer for initiation of replication of ColE1 DNA by ribonuclease H. *Proc. Natl. Acad. Sci. USA*, **77**, 2450–2454.

Jacob, F., Brenner, S. & Cuzin, F. (1963). On the regulation of DNA replication in bacteria. *Cold Spring Harbor Symp. Quant. Biol.*, **28**, 329–348.

Jacoby, G. A. (1974). *Antimicrob. Agents Chemother.*, **6**, 807–810.

Jaffe, A., Ogura, T. & Hiraga, S. (1985). Effects of the ccd function of the F plasmid on bacterial growth. *J. Bacteriol.*, **163**, 841–849.

James, A. A. & Kolodner, R. (1983). Genetic recombination of plasmids in *Escherichia coli*. In *Mechanisms of DNA Replication and Recombination* (pp. 761–772). New York: Alan R. Liss.

James, A. A., Morrison, P. T. & Kolodner, R. (1982). Genetic recombination of bacterial plasmid DNA. Analysis of the effect of recombination-deficient mutations on plasmid recombination. *J. Molec. Biol.*, **160**, 411–430.

James, A. A., Morrison, P. T. & Kolodner, R. (1983). Isolation of genetic elements that increase frequencies of plasmid recombinants. *Nature*, **303**, 256–259.

Johnson, R. & Adams, J. (1992). The ecology and evolution of tetracycline resistance. *TREE*, **7**, 295–299.

Jones, I. M., Primrose, S. B., Robinson, A. & Ellwood, D. C. (1980). Maintenance of some ColE1-type plasmids in chemostat culture. *Molec. Gen. Genet.*, **180**, 579–584.

Jones, R. T. & Curtiss, R. (1970). Genetic exchange between *Escherichia coli* strains in the mouse intestine. *J. Bacteriol.*, **103**, 71–80.

Kasuya, M. (1964). Transfer of drug resistance between enteric bacteria induced in the mouse intestine. *J. Bacteriol.*, **88**, 322–328.

Keasling, J. D., Palsson, B. O. & Cooper, S. (1992). Replication of prophage P1 is cell-cycle specific. *J. Bacteriol.*, **173**, 4457–4462.

Kieser, T., Hopwood, D. A., Wright, H. M. & Thompson, C. J. (1982). plJ101, a multi-copy broad host-range *Streptomyces* plasmid: functional analysis and development of DNA cloning vectors. *Molec. Gen. Genet.*, **185**, 223–238.

Kinashi, H. & Shimaji-Murayama, M. (1991). Physical characterization of SCP1, a giant linear plasmid from *Streptomyces coelicolor*. *J. Bacteriol.*, **173**, 1523–1529.

Kinashi, H., Shimaji, M. & Sakai, A. (1987). Giant plasmids in *Streptomyces* which code for antibiotic biosynthesis genes. *Nature*, **328**, 454–456.

Kinashi, H., Murayama, M., Matsushita, H. & Nimi, O. (1993). Structural analysis of the giant linear plasmid SCP1 in various *Streptomyces coelicolor* strains. *J. Gen. Microbiol.*, **139**, 1261–1269.

Kittel, B. L. & Helinski, D. R. (1991). Iteron inhibition of plasmid, RK2 replication *in vitro*: evidence for intermolecular coupling of replication origins as a mechanism for RK2 replication control. *Proc. Natl. Acad. Sci.*, **88**, 1389–1393.

Koepsel, R. R., Murray, R. W. & Kahn, S. A. (1986). Sequence-specific interactions between the replication initiator protein of plasmid pT181 and its origin of replication. *Proc. Natl. Acad. Sci. USA*, **83**, 5484–5488.

Kogoma, T. (1984). Absence of RNase H allows replication of pBR322 in

Escherichia coli mutants lacking DNA polymerase I. *Proc. Natl. Acad. Sci. USA*, **81**, 7845–7849.

Kokjohn, T. A. & Miller, R. V. (1992). Gene transfer in the environment: transduction. In *Release of genetically engineered and other micro-organisms* (eds. J. C. Fry, M. J. Day), pp. 54–81. Cambridge, UK: Cambridge University Press.

Kollek, R., Oertel, W. & Goebel, W. (1978). Isolation and characterization of the minimal fragment required for autonomous replication of a copy mutant pKN102 of the antibiotic resistance factor R1. *Molec. Gen. Genet.*, **162**, 51–58.

Kolodner, R., Fishel, R. A. & Howard, M. (1985). Genetic recombination of bacterial plasmid DNA: effect of RecF pathway mutations on plasmid recombination in *Escherichia coli*. *J. Bacteriol.*, **163**, 1060–1066.

Kolot, M. (1990). Par site of the ColN plasmid: structural and functional organization. *Molec. Gen. Genet.*, **222**, 77–80.

Kolter, R. (1981). Replication properties of plasmid R6K. *Plasmid*, **5**, 2–9.

Kornberg, A. (1982). *Supplement to DNA Replication*. San Francisco: W. H. Freeman.

Kowalski, D. & Eddy, M. J. (1989). The DNA unwinding element: a novel, *cis*-acting component that facilitates opening of the *Escherichia coli* replication origin. *EMBO J.*, **8**, 4335–4344.

Kratz, J., Schmidt, F. & Wiedemann, B. (1983). Characterization of Tn*2411* and Tn*2410*, two transposons derived from R-plasmid R1767 and related to Tn*2603* and Tn*21*. *J. Bacteriol.*, **155**, 1333–1342.

Krause, M. & Guiney, D. G. (1991). Identification of a multimer resolution system involved in stabilization of the *Salmonella dublin* virulence plasmid pSDL2. *J. Bacteriol.*, **173**, 5754–5762.

Krause, M, Fang, F. C. & Guiney, D. G. (1992). Regulation of plasmid virulence gene expression in *Salmonella dublin* involves an unusual operon structure. *J. Bacteriol.*, **174**, 4482–4489.

Kunz, I. D. (1992). Structure-based strategies for drug design and discovery. *Science*, **257**, 1078–1082.

Laban, A. & Cohen, A. (1981) Interplasmidic and intraplasmidic recombination in *Escherichia coli* K-12 *Molec. Gen. Genet.*, **184**, 200–207.

Lacks, S. A., Lopez, P., Greenberg, B. & Espinosa, M. (1986). Identification and analysis of genes for tetracycline resistance and replication functions in the broad-host-range plasmid pLS1. *J. Molec. Biol.*, **192**, 753–765.

Lane, D., Rothenbuehler, R., Merrillat, A.-M. & Aiken, C. (1987). Analysis of the F plasmid centromere. *Molec. Gen. Genet.*, **207**, 406–412.

Lebek, G. (1963). *Zentralbl. Bakteriol, Parasitenkd, Infektionskr Hyg, Abt 1: Orig, Reihe A*, **188**, 494–505.

Lederberg, J. (1952). Cell genetics and hereditary symbiosis. *Physiol. Rev.*, **32**, 403–430.

Lederberg, J. & Tatum, E. L. (1946). Gene recombination in *E. coli*. *Nature*, **158**, 558.

Lederberg, J., Lederberg, E. M., Zinder, N. D. & Lively, E. R. (1951). Recombination analysis of bacterial heredity. *Cold Spring Harbor Symp. Quant. Biol.*, **16**, 413–443.

Lederberg, J., Cavalli, L. L. & Lederberg, E. M. (1952). Sex compatibility in *E. coli*. *Genetics*, **37**, 720–730.

Lee, S. B. & Bailey, J. E. (1984a). A mathematical model for lambda-dv plasmid replication: analysis of wild-type plasmid. *Plasmid*, **11**, 151–165.

Lee, S B. & Bailey, J. E. (1984b). A mathematical for lambda-dv plasmid replication: analysis of copy number mutants. *Plasmid*, **11**, 166–177.

Lee, S. H., Frost, L. S. & Paranchych, W. (1992). FinOP repression of the F plasmid involves extension of the half-life of FinP antisense RNA by FinO. *Molec. Gen. Genet.*, **235**, 131–139.

References

Lehnherr, H., Maguin, E., Jafri, S. & Yarmolinsky, M. B. (1993). Plasmid addiction genes of bacteriophage P1: *doc*, which causes cell death on curing of prophage, and *phd*, which prevents host death when prophage is retained. *J. Molec. Biol.*, **233**, 414–428.

Lessl, M., Balzer, D., Lurz, R., Waters, V., Guiney, D. & Lanke, E. (1992). Dissection of IncP conjugative plasmid transfer: Definition of the transfer region Tra2 by mobilization of the Tra1 region in *trans*. *J. Bacteriol.*, **174**, 2493–2500.

Lessl, M., Balzer, D., Weyrauch, K. & Lanke, E. (1993). The mating pair formation system of plasmid RP4 defined by RSF1010 mobilization and donor-specific phage propagation. *J.Bacteriol.*, **175**, 6415–6425.

Lett, M. C., Bennett, P. M. & Vidon, D. J. M. (1985). Characterization of Tn*3926*, a new mercury-resistance transposon from *Yersinia enterosolitica*. *Gene*, **40**, 79–91.

Levin, B. R. (1988). Frequency-dependent selection in bacterial populations. *Phil. Trans. R. Soc. Lond. B*, **319**, 459–472.

Levin, B. R. & Lenski, R. E. (1983). Coevolution in Bacteria and Their Viruses and Plasmids. In *Coevolution* (ed. D. J. Slatkin & M. Futyma), pp. 99–127. Sunderland, MA.: Sinauer Associates.

Levy, S. B. (1984). Resistance to the Tetracyclines. In *Antimicrobial Drug Resistance* (ed. L. E. Bryan), pp. 191–240. London: Academic Press.

Levy, S. B. (1986). Ecology of Antibiotic Resistance Determinants. In *Antibiotic resistance genes: ecology, transfer and expression* (eds. P. Novick & S. B. Levy), pp. 17–30. New York: Cold Spring Harbor Laboratory.

Light, J. & Molin, S. (1982). The sites of action of the two copy number control functions of plasmid R1. *Molec. Gen. Genet.*, **187**, 486–493.

Lin-Chao, S. & Bremer, H. (1986). Effect of the bacterial growth rate on replication control of plasmid pBR322 in *Escherichia coli*. *Molec. Gen. Genet.*, **203**, 143–149.

Lin-Chao, S. & Bremer, H. (1987). Activities of the RNA I and RNA II promoters of plasmid pBR322. *J. Bacteriol.*, **169**, 1217–1222.

Lin-Chao, S. & Cohen, S. N. (1991). The rate of processing and degradation of antisense RNA I regulates the replication of ColE1-type plasmids *in vivo*. *Cell*, **65**, 1233–1242.

Lloyd, R. G., Picksley, S. M. & Prescott, C. (1983). Inducible expression of a gene specific to the RecF pathway for recombination in *Escherichia coli* K-12. *Molec. Gen. Genet.*, **190**, 162–167.

Lloyd, R. G., Benson, F. E. & Shurvington, C. E. (1984). Effect of *ruv* mutations on recombination and DNA repair in *Escherichia coli* K-12. *Molec. Gen. Genet.*, **194**, 303–309.

Loh, S. M., Cram, D. S. & Skurray, R. A. (1988). Nucleotide sequence and transcriptional analysis of a third function Flm involved in F plasmid maintenance. *Gene*, **66**, 259–268.

Lopez, J., Delgado, D., Andreis, I., Ortiz, J. M. & Rodriguez, J. C. (1991). Isolation and evolutionary analysis of a RepFVIB replicon of the plasmid pSU212. *J. Gen. Microbiol.*, **137**, 1093–1099.

Lovett, S. T. & Clark, A. J. (1984). Genetic analysis of the *recJ* gene of *Escherichia coli* K12. *J. Bacteriol.*, **157**, 190–196.

Lundquist, P. D. & Levin, B. R. (1986). Transitory derepression and the maintenance of conjugative plasmids. *Genetics*, **113**, 483–497.

Luria, S. E. & Suit, J. L. (1987). Colicins and Col plasmids. In J. Ingraham, K. Brooks Low, B. Magasanik, M. Schaechter, H. E. Umbarger, & F. C. Neidhardt (eds), *Escherichia coli and Salmonella typhimurium: Cellular and Molecular Biology* (pp. 1615–1624). Washington: ASM Press.

Luttinger, A. (1995). The twisted life of DNA in the cell—bacterial topoisomerases. *Molec. Microbiol.*, **15**, 601–606.

Lyon, B. R. & Skurray, R. (1987). Antimicrobial resistance of *Staphylococcus aureus*: genetic basis. *Microbiol. Rev.*, **51**, 88–134.

Maas, R., Saadi, S. & Maas, W. K. (1989). Properties and incompatibility behaviour of miniplasmids derived from the bireplicon plasmid pCG86. *Molec. Gen. Genet.*, **218**, 190–198.

Maclean, I. H., Rogers, K. B. & Fleming, A. (1939). M. and B. 693 and Pneumococci. *Lancet*, **i**, 562–568.

Maneewannakul, S., Maneewannakul, K. & Ippen-lhler, K. (1992a). Characterization, localization, and sequence of F transfer region products: the pilus assembly gene product TraW and a new product, Trbl. *J. Bacteriol.*, **174**, 5567–5574.

Maneewannakul, S., Kathir, P. & Ippen-lhler, K. (1992b). Characterization of the F plasmid mating aggregation gene *traN* and of a new F transfer region locus *trbE*. *J. Molec. Biol.*, **225**, 299–311.

Manen, D., Goebel, T. & Caro, L. (1990). The *par* region of pSC101 affects plasmid copy number as well as stability. *Molec. Microbiol.*, **4**, 1839–1846.

Marmur, J., Rownd, R., Falkow, S., Baron, L. S., Schildkraut, C. & Doty, P. (1961). The nature of intergeneric episomal infection. *Proc. Natl. Acad. Sci.*, **47**, 972–979.

Martin, K. A., Friedman, S. A. & Austin, S. J. (1987). Partition site of the P1 plasmid. *Proc. Natl. Acad. Sci. USA*, **84**, 8544–8547.

Martinez, E. & de la Cruz, F. (1988). Transposon Tn21 encodes a RecA-independent site-specific integration system. *Molec. Gen. Genet.*, **211**, 320–325.

Martinez, E. & de la Cruz, F. (1990). Genetic elements involved in Tn21 site-specific integration, a novel mechanism for the dissemination of antibiotic resistance genes. *EMBO J.*, **9**, 1275–1281.

Masukata, H. & Tomizawa, J. (1984). Effects of point mutations on formation and structure of the RNA primer for ColE1 DNA replication. *Cell*, **36**, 513–522.

Masukata, H., Dasgupta, S. & Tomizawa, J. (1987). Transcriptional activation of ColE1 DNA synthesis by displacement of the nontranscribed strand. *Cell*, **51**, 1123–1130.

Masukata, H. & Tomizawa, J. (1990). A mechanism of formation of a persistent hybrid between elongating RNA and template DNA. *Cell*, **62**, 331–338.

Matsubara, K. (1976). Genetic structure and regulation of a replicon of plasmid lambda-dv. *J. Molec. Biol.*, **102**, 427–439.

Mazaitis, A. J., Maas, R. & Maas, W. K. (1981). Structure of a naturally occurring plasmid with genes for enterotoxin production and drug resistance. *J. Bacteriol.*, **145**, 97–105.

Mazodier, P. & Davies, J. (1991). Gene transfer between distantly related bacteria. *Ann. Rev. Biochem.*, **25**, 147–171.

McEachern, M. J., Bott, M. A., Tooker, P. A. & Helinski, D. R. (1989). Negative control of plasmid R6K replication: possible role of intermolecular coupling of replication origins. *Proc. Acad. Sci.*, **86**, 7942–7946.

Meacock, P. A. & Cohen, S. N. (1980). Partitioning of bacterial plasmids during cell division: a *cis*-acting locus that accomplishes stable inheritance. *Cell*, **20**, 529–542.

Meinhardt, F., Kempken, F., Kamper, F. & Esser, K. (1990). Linear plasmids among eukaryotes: fundamentals and application. *Curr. Genet.* **17**, 89–95.

Mergeay, M., Lejeune, P., Sadouk, A., Gerits, J. & Fabry, L. (1987). Shuttle transfer (or retrotransfer) of chromosomal markers mediated by plasmid pULB113. *Molec. Gen. Genet.*, **209**, 61–70.

Meyer, J. F., Nies, B. A. & Wiedemann, B. (1983). Amikacin resistance mediated by multiresistance transposon Tn2424. *J. Bacteriol.*, **155**, 755–760.

Meynell, E., Meynell, G. G. & Datta, N. (1968). Phylogenetic relationships of drug-resistance factors and other transmissible plasmids. *Bacter Rev*, **32**, 5–83.

Michiels, T. & Cornelis, G. (1984). Detection and characterization of Tn*2501*, a transposon included within the lactose transposon Tn*951*. *J. Bacteriol.*, **158**, 866–871.

Miki, T., Yoshioka, K. & Horiuchi, T. (1984). Control of cell division by sex factor F in *Escherichia coli*. I. The 42.84–43.6 F segment couples cell division of the host bacteria with replication of plasmid DNA. *J. Molec. Biol.*, **174**, 605–625.

Miki, T., Park, J. A., Nagao, K., Murayama, N. & Horiuchi, T. (1992). Control of segregation of chromosomal DNA by sex factor F in *Escherichia coli*. Mutants of DNA gyrase subunit A suppress *letD* (*ccdB*) product growth inhibition. *J. Molec. Biol.*, **225**, 39–52.

Miller, C. A., Beaucage, S. L. & Cohen, S. N. (1990). Role of DNA superhelicity in partitioning of the pSC101 plasmid. *Cell*, **62**, 127–133.

Min, Y.-N., Tabuchi, A., Fan, Y.-L., Womble, D. D. & Rownd, R. H. (1988). Complementation of mutants of the stability locus of IncFII plasmid NR1. Essential functions of the *trans*-acting *stbA* and *stbB* gene products. *J. Molec. Biol.*, **204**, 345–356.

Minton, N. P., Oultram, J. D., Brehm, J. K. & Atkinson, T. (1988). The replication proteins of plasmids pE194 and pLS1 have N-terminal homology. *Nucl. Acids Res.*, **16**, 3101.

Mitsuhashi, S., Harada, K. & Hashimoto, H. (1960a). Multiple resistance of enteric bacteria and transmission of drug-resistance to other strain by mixed cultivation. *Japan. J. Exper. Med.*, **30**, 179–184.

Mitsuhashi, S., Hashimoto, H., Harada, K., Suzuki, M., Kameda, M. & Matsuyama, T. (1960b). Multiple resistance of *S. flexneri* 3a and *E. coli* isolated from the epidemy in Gunma Prefecture. *Japan. J. Bacteriol.*, **15**, 844–848 (in Japanese).

Mitsuhashi, S., Harada, K., Hashimoto, H. & Egawa, R. (1961). Drug resistance of enteric bacteria. 5. Drug-resistance of *Escherichia coli* isolated from human being. *Japan J. Exper. Med.*, **31**, 53–60.

Mitsuhashi, S. (1971). Epidemiology of Bacterial Drug Resistance. In *Transferable Drug Resistance Factor R* (ed. S. Mitsuhashi), pp. 1–23. Baltimore: University Park Press.

Molin, S. & Nordström, K. (1980). Control of plasmid R1 replication: functions involved in replication, copy number control, incompatibility, and switch-off of replication. *J. Bacteriol.*, **141**, 111–120.

Morgenroth, J. & Kaufmann M. (1912) Z. *Immunitaetsforsch.*, **15**, 610.

Mori, H., Kondo, A., Ohshima, A., Ogura, T. & Hiraga, S. (1986). Structure and function of the F plasmid genes essential for partitioning. *J. Molec. Biol.*, **192**, 1–15.

Morlon, J., Chartier, M., Bidaud, M. & Lazdunski, C. (1988). The complete nucleotide sequence of the colicinogenic plasmid ColA. High extent of homology with ColE1. *Molec. Gen. Genet.*, **211**, 231–243.

Motallebi-Veshareh, M., Rouch, D. A. & Thomas, C. M. (1990). A family of ATPases involved in active partitioning of diverse bacterial plasmids. *Molec. Microbiol.*, **4**, 1455–1463.

Muraiso, K., Mukhopadhyay, G. & Chattoraj, D. K. (1990). Location of a P1 plasmid replication inhibitor determinant within the initiator gene. *J. Bacteriol*, **172**, 4441–4447.

Murotsu, T. & Matsubara, K. (1980). Role of an autorepression system in the control of lambda dv plasmid copy number and incompatibility. *Molec. Gen. Genet.*, **179**, 509–519.

Naito, S., Kitani, T., Ogawa, T., Okazaki, T. & Uchida, H. (1984). *Escherichia coli* mutants suppressing replication-defective mutations of the ColE1 plasmid. *Proc. Natl. Acad. Sci. USA*, **81**, 550–554.

Nakayama, H., Nakayama, K., Nakayama, R., Irino, N., Nakayama, Y. &

Hanawalt, P. C. (1984). Isolation and genetic characterization of a thymineless death-resistant mutant of *Escherichia coli* K-12: identification of a new mutation *recQ1* that blocks the RecF recombination pathway. *Molec. Gen. Genet.*, **195**, 474–480.

Neill & Holmes (1988). Genetics of Toxinogenic Bacteria. In *Bactorial toxins* (ed. M. C. Hardegree & A. T. Tu), pp. 383–416. New York: Marcel Dekker.

Neu, H. C. (1992). The crisis in antibiotic resistance. *Science*, **257**, 1064–1073.

Nielsen, P. F. & Molin, S. (1984). How the R1 replication control system responds to copy number deviations. *Plasmid*, **11**, 264–267.

Nies, B. A., Meyer, J. F. & Wiedemann, B. (1986). Role of transposition and homologous recombination in the rearrangement of plasmid DNA. *J. Antimicrob. Chemother.*, **18**, 35–41.

Nordström, K., Ingram, L. C. & Lundback, A. (1972). Mutations in R-factors of *Escherichia coli* causing an increased number of R-factor copies per chromosome. *J. Bacteriol.*, **110**, 562–569.

Nordström, K., Molin, S. & Aagaard-Hansen, H. (1980a). Partitioning of plasmid R1 in *Escherichia coli*. I. Kinetics of loss of plasmid derivatives deleted of the par region. *Plasmid*, **4**, 215–227.

Nordström, K., Molin, S. & Aagaard-Hansen, H. (1980b). Partitioning of plasmid R1 in *Escherichia coli*. II. Incompatibility properties of the partitioning system. *Plasmid*, **4**, 332–349.

Nordström, K. & Aagaard-Hansen, H. (1984). Maintenance of bacterial plasmids: comparison of theoretical calculations and experiments with plasmid R1 in *Escherichia coli*. *Molec. Gen. Genet.*, **197**, 1–7.

Nordström, K., Molin, S. & Light, J. (1984). Control of replication of bacterial plasmids: genetics, molecular biology, and physiology of the plasmid R1 system. *Plasmid*, **12**, 71–90.

Nordström, K. (1985). Control of plasmid replication: theoretical considerations and practical solutions. In D. R. Helinski, S. N. Cohen, D. B. Clewell, D. A. Jackson, & A. Hollaender (eds), *Plasmids in Bacteria* (pp. 189–214). New York: Plenum Press.

Nordström, K. & Austin, S. J. (1989). Mechanisms that contribute to the stable segregation of plasmids. *Ann. Rev. Genet.*, **23**, 37–69.

Nordström, K. (1990). Control of plasmid replication—how do DNA iterons set the replication frequency? *Cell*, **63**, 1121–1124.

Nordström, K. & Austin, S. J. (1993). Cell-cycle-specific initiation of replication. *Molec. Microbiol.*, **10**, 457–463.

Novick, R. P., Edelman, I. & Lofdahl, S. (1986). Small *Staphylococcus aureus* plasmids are transduced as linear multimers that are formed and resolved by replicative processes. *J. Molec. Biol.*, **192**, 209–220.

Novick, R. P. (1989). Staphylococcal plasmids and their replication. *Ann. Rev. Microbiol.*, **43**, 537–565.

Novick, R. P., Iordanescu, S., Projan, S., Kornblum, J. & Edelman, I. (1989). pT181 plasmid replication is regulated by a countertranscript-driven transcriptional attenuator. *Cell*, **59**, 395–404.

O'Connor, M. B., Kilbane, J. J. & Malamy, M. H. (1986). Site-specific and illegitimate recombination in the *oriV1* region of the F factor. DNA sequences involved in recombination and resolution. *J. Molec. Biol.*, **189**, 85–102.

Ochman, H. & Wilson, A. C. (1987). Evolutionary History of Enteric Bacteria. In *Escherichia coli and Salmonella typhimurium cellular and molecular biology* (eds. F. C. Neidhardt), pp. 1649–1654. Washington DC: American Society for Microbiology.

Ogura, T. & Hiraga, S. (1983a). Partition mechanism of F plasmid: two plasmid gene-encoded products and a *cis*-acting region are involved in partition. *Cell*, **32**, 351–360.

Ogura, T. & Hiraga, S. (1983b). Mini-F plasmid genes that couple host cell division to plasmid proliferation. *Proc. Natl. Acad. Sci. USA*, **80**, 4784–4788.

Ogura, T., Niki, H., Mori, H., Morita, M., Hasegawa, M., Ichinose, C. & Hiraga, S. (1990). Identification and characterization of *gyrB* mutants of *Escherichia coli* that are defective in partitioning of mini-F plasmids. *J. Bacteriol.*, **172**, 1562–1568.

Öhman, M. & Wagner, E. G. H. (1989). Secondary structure analysis of the RepA mRNA leader transcript involved in the control of replication of plasmid R1. *Nucl. Acids Res.*, **17**, 2557–2579.

Pal, S. K. & Chattoraj, D. K. (1988). P1 plasmid replication: initiator sequestration is inadequate to explain control by initiator-binding sites. *J. Bacteriol*, **170**, 3554–3560.

Pal, S. K., Mason, R. J. & Chattoraj, D. K. (1986). P1 plasmid replication: role of initiator titration in copy number control. *J. Molec. Biol.*, **192**, 275–285.

Pansegrau, W., Schröder, W. & Lanka, E. (1993). Relaxase (TraI) of IncPα plasmid RP4 catalyzes a site-specific cleaving–joining reaction of single-stranded DNA. *Proc. Natl. Acad. Sci. USA*, **90**, 2925–2929.

Patient, M. E. & Summers, D. K. (1993). ColE1 multimer formation triggers inhibition of *E. coli* cell division. *Molec. Microbiol.*, **8**, 1089–1095.

Perea, E. J., Daza, R. M. & Mendaza, M. P. (1977). *Chemotherapy (Basel)*, **23** (Suppl. 1), 127–132.

Persson, C., Wagner, E. G. H. & Nordström, K. (1988). Control of replication of plasmid R1: kinetics of *in vitro* interaction between the antisense RNA, CopA, and its target, CopT. *EMBO J.*, **7**, 3279–3288.

Persson, C., Wagner, E. G. H. & Nordström, K. (1990). Control of replication of plasmid R1: formation of an initial transient complex is rate-limiting for antisense RNA–target RNA pairing. *EMBO J.*, **9**, 3777–3785.

Petrocheilou, V., Grinsted, J. & Richmond, M. H. (1976). R plasmid transfer *in vivo* in the absence of antibiotic selection pressure. *Antimicrob. Agents Chemother.*, **10**, 753–761.

Pettis, G. S. & Cohen, S. N. (1994). Transfer of the plJ101 plasmid in *Streptomyces lividans* requires a *cis*-acting function dispensible for chromosomal gene transfer. *Molec. Microbiol.*, **13**, 955–964.

Plasterk, R. H. A., Simon, M. I. & Barbour, A. G. (1985). Transposition of structural genes to an expression sequence on a linear plasmid causes antigenic variation in the bacterium *Borrelia hermsii*. *Nature*, **318**, 257–263.

Polisky, B. (1988). ColE1 replication control circuitry: sense from antisense. *Cell*, **55**, 929–932.

Poulsen, L. K., Larsen, N. W., Molin, S. & Andersson, P. (1989). A family of genes encoding a cell-killing function may be conserved in all Gram-negative bacteria. *Molec. Microbiol.*, **3**, 1463–1472.

Poyart-Salmeron, C., Trieucuot, P., Carlier, C. & Courvalin, P. (1990). The integration-excision system of the conjugative transposon Tn1545 is structurally and functionally related to those of lamboid phages. *Molec. Microbiol.*, **4**, 1513–1521.

Pritchard, R., Barth, P. T. & Collins, J. (1969). Control of DNA synthesis in bacteria. *Symp. Soc. Gen. Microbiol.*, **19**, 263–297.

Pritchard, R. H. (1978). Control of DNA replication in bacteria. In I. Molineux & M. Kohiyama (eds), *DNA Synthesis: Present and Future* (pp. 1–26). New York: Plenum Press.

Pritchard, R. H. (1984). Control of DNA replication in bacteria. In P. Nurse & E. Streiblova (eds), *The Microbial Cell Cycle* (pp. 19–27). Boca Raton, FL: CRC Press.

Projan, S. J. & Moghazeh, S. L. (1991). Termination of replication and its potential role in cassette assembly of the single-stranded plasmids of Gram-positive bacteria. *Plasmid*, **25**, 241.

Projan, S. J. & Novick, R. P. (1992). *Cis*-inhibitory elements in the pT181 replication system. *Plasmid*, **27**, 81–92.

Rafii, F. & Crawford, D. L. (1989). Gene Transfer Among Streptomyces. In *Gene Transfer in the Environment* (eds. S. B. Levy & R. V. Miller), pp. 309–355. New York: McGraw-Hill.

Reanney, D. (1976). Extrachromosomal elements as possible agents of adaptation and development. *Bacter. Rev.*, **40**, 552–590.

Recchia, G. D., Stokes, H. W. & Hall, R. M (1994). Characterization of specific and secondary recombination sites recognized by the integron DNA integrase. *Nucl. Acids Res.*, **22**, 2071–2078.

Recchia, G. D. & Hall, R. M. (1995). Plasmid evolution by acquisition of mobile gene cassettes: plasmid pIE723 contains the *aadB* gene cassette precisely inserted at a secondary site in the IncQ plasmid RSF1010. *Molec. Microbiol.*, **15**, 179–187.

Reece, R. J. & Maxwell, A. (1991). DNA gyrase—structure and function. *Crit. Rev. Biochem. Molec. Biol.*, **26**, 335–375.

Reed, N. D., Sieckmann, D. G. & Georgi, C. E. (1969). Transfer of infectious drug resistance in microbially defined mice. *J. Bacteriol.*, **100**, 22–26.

Riise, E., Stougaard, P., Bindslev, B., Nordström, K. & Molin, S. (1982). Molecular cloning and characterization of a copy number control gene *copB* of plasmid R1. *J. Bacteriol.*, **151**, 1136–1145.

Roberts, R. C., Burioni, R. & Helinski, D. R. (1990). Genetic characterization of the stabilizing functions of a region of broad-host-range plasmid RK2. *J. Bacteriol.*, **172**, 6204–6216.

Rownd, R. H., Womble, D. D., Dong, X., Luckow, V. A. & Wu, R. P. (1985). Incompatibility and IncFII plasmid replication control. In D. R. Helinski, S. N. Cohen, D. B. Clewell, D. A. Jackson, & A. Hollander (eds), *Plasmids in Bacteria* (pp. 335–354). New York: Plenum Press.

Ruby, C. & Novick, R. P. (1975). Plasmid interactions in *Staphylococcus aureus*: nonadditivity of compatible plasmid DNA pools. *Proc. Natl. Acad. Sci. USA*, **72**, 5031–5035.

Ruiz-Echevarria, M. J., de Torrontegui, G., Gimenez-Gallego, G. & Diaz-Orejas, R. (1991). Structural and functional comparison between the stability systems ParD of plasmid R1 and Ccd of plasmid F. *Molec. Gen. Genet.*, **225**, 355–362.

Saadi, S., Maas, W. K., Hill, D. F. & Bergquist, P. L. (1987). Nucleotide sequence analysis of RepFIC, a basic replicon in IncFI plasmids P307 and F and its relation to the RepA replicon of IncFII plasmids. *J. Bacteriol.*, **164**, 1836–1846.

Sakaguchi, K., Hirochika, H. & Gunge, N. (1985). Linear Plasmids with Terminal Inverted Repeats Obtained from *Streptomyces rochei* and *Kluyveromyces*. In *Plasmids in Bacteria* (eds. D. R. Helinski, S. N. Cohen, D. B. Clewell, D. A. Jackson & A. Hollaender), pp. 433–451. New York: Plenum Press.

Sambrook, J., Fritsch, E. F. & Maniatis, T. (1989). *Molecular Cloning. A Laboratory Manual*. New York: Cold Spring Harbor Laboratory Press.

Saunders, J. R., Hart, C. A. & Saunders, V. A. (1986). Plasmid-mediated resistance to β-lactam antibiotics in Gram-negative bacteria: the role of *in-vivo* recyclization reactions in plasmid evolution. *J. Antimicrob. Chemother.*, **18**, 57–66.

Saye, D. J., Kokjohn, T. A. & Miller, R. V. (1987). *Pseudomonas aeruginosa* Phage DS1 Suppresses the Les- Phenotype. In *Abstracts of the annual meetings of the American Society for Microbiology*, p. 238. Washington DC: American Society for Microbiology.

Schmidt, F. (1984). The role of insertions, deletions and substitutions in the evolution of R6 related plasmids encoding aminoglycoside transferase ANT-(2''). *Molec. Gen. Genet.*, **194**, 248–259.

Schmitt, R., Bernhard, E. & Mattes, R. (1979). Characterization of Tn*1721*, a new transposon containing tetracycline resistance genes capable of amplification. *Molec. Gen. Genet.*, **172**, 53–65.

Schwartz, D. C. & Cantor, C. R. (1984). Separation of yeast chromosomal-sized DNAs by pulsed field gradient gel electrophoresis. *Cell*, **37**, 67–75.

Scott, J. R. (1992). Sex and the single circle: conjugative transposition. *J. Bacteriol.*, **174**, 6005–6010.

Scott, J. R., Bringel, F., Marra, D., Van Aistine, G. & Rudy, C. K. (1994). Conjugative transposition of Tn916: preferred targets and evidence for conjugative transfer of a single strand and for a double-stranded circular intermediate. *Molec. Microbiol.*, **11**, 1099–1108.

Selzer, G., Som, T., Itoh, T. & Tomizawa, J. (1983). The origin of replication of plasmid p15A and comparative studies on the nucleotide sequences around the origin of related plasmids. *Cell*, **32**, 119–129.

Simonsen, L. (1991). The existence conditions for bacterial plasmids: theory and reality. *Microb. Ecol.*, **22**, 187–205.

Smith, G. R. (1988). Homologous recombination in prokaryotes. *Microbiol. Rev.*, **52**, 1–28.

Sompayrac, L. & Maaloe, O. (1973). Autorepressor model for control of DNA replication. *Nature New Biology*, **241**, 133–135.

Stahl, F. W. (1979). Special sites in generalized recombination. *Ann. Rev. Genet.*, **13**, 7–24.

Stanish, V. A. (1988). Identification and analysis of plasmids at the genetic level. In *Methods in Microbiology*, **2**(1), pp. 11–47.

Sternberg, N. & Austin, S. (1983). Isolation and characterization of P1 minireplicons, lambda-P1: 5R and lambda-P1: 5L. *J. Bacteriol*, **153**, 800–812.

Stewart, G. J. (1992). Gene Transfer in the environment: transformation. In *Release of genetically engineered and other micro-organisms* (eds. J. C. Fry and M. J. Day). Cambridge, UK: Cambridge University Press.

Stewart, F. M. & Levin, B. R. (1977). The population biology of bacterial plasmids: *a priori* conditions for the existence of conjugationally transmitted factors. *Genetics*, **87**, 209–228.

Stirling, C. J., Colloms, S. D., Collins, J. F., Szatmari, G. & Sherratt, D. J. (1989). An *Escherichia coli* gene required for plasmid ColE1 site-specific recombination is identical to *pepA*, encoding aminopeptidase A, a protein with substantial similarity to bovine lens leucine aminopeptidase. *EMBO J.*, **8**, 1623–1627.

Stirling, C. J., Szatmari, G., Stewart, G., Smith, M. C. M. & Sherratt, D. J. (1988). The arginine repressor is essential for plasmid stabilizing site-specific recombination at the ColE1 *cer* locus. *EMBO J.*, **7**, 4389–4395.

Stokes, H. W. & Hall, R. M. (1989). A novel family of potentially mobile DNA elements encoding site-specific gene-integration functions: integrons. *Molec. Microbiol.*, **3**, 1669–1683.

Stougaard, P., Molin, S. & Nordström, K. (1981). RNAs involved in copy number control and incompatibility of plasmid R1. *Proc. Natl. Acad. Sci. USA.*, **78**, 6008–6012.

Summers, D. K. (1989). Derivatives of ColE1 *cer* show altered topological specificity in site-specific recombination. *EMBO J.*, **8**, 309–315.

Summers, D. K. & Sherratt, D. J. (1984). Multimerization of high copy number plasmids causes instability: ColE1 encodes a determinant essential for plasmid monomerization and stability. *Cell*, **36**, 1097–1103.

Summers, D. K. & Sherratt, D. J. (1985). Bacterial plasmid stability. *Bioessays*, **2**, 209–211.

Summers, D. K., Beton, C. W. H. & Withers, H. L. (1993). Multicopy plasmid instability: the dimer catastrophe hypothesis. *Molec. Microbiol.*, **8**, 1031–1038.

Summers, D., Yaish, S., Archer, J. & Sherratt, D. (1985). Multimer resolution systems of ColE1 and ColK: localisation of the crossover site. *Molec. Gen. Genet.*, **201**, 334–338.

Swack, J. A., Pal, S. K., Mason, R. J., Abeles, A. L. & Chattoraj, D. K. (1987). P1 plasmid replication: measurement of the initiator protein concentration *in vivo*. *J. Bacteriol.*, **169**, 3737–3742.

Swann (1969). *Joint Committee on the Use of Antibiotics in Animal and Veterinary Medicine* (No. 4190). HM Stationary Office, London.

Tabuchi, A., Min, Y.-N., Kim, C. K., Fan, Y.-L., Womble, D. D. & Rownd, R. H. (1988). Genetic organizations and nucleotide sequence of the stability locus of the incFII plasmid NR1. *J. Molec. Biol.*, **202**, 511–525.

Tam, J. E. & Kline, B. C. (1989). The F plasmid ccd autorepressor is a complex of CcdA and CcdB proteins. *Mol. Gen. Genet.*, **219**, 26–32.

Tam, J. & Polisky, B. (1983). Structural analysis of RNA molecules involved in plasmid copy number control. *Nucl. Acids Res.*, **11**, 6381–6397.

Tanimoto, K. & Clewell, D. B. (1993). Regulation of the pAD1-encoded sex pheromone response in *Enterococcus faecalis*: expression of the positive regulator TraE1. *J. Bacteriol.*, **175**, 1008–1018.

Tauxe, R. V., Cavanagh, T. R. & Cohen, M. L. (1989). Interspecies gene transfer *in vivo* producing an outbreak of multiply resistant shigellosis. *J. Infect. Dis.*, **160**, 1067–1070.

Thistead, T. & Gerdes, K. (1992). Mechanism of post-segregational killing by the *hok/sok* system of plasmid R1. *Sok* antisense RNA regulates *hok* gene expression indirectly through the overlapping *mok* gene. *J. Mol. Boil,*, **223**, 41–54.

Thomas, C. M. (1988). Recent studies on the control of plasmid replication. *Biochim. Biophys. Acta*, **949**, 253–263.

Thomas, C. M. & Helinski, D. R. (1989). Vegetative Replication and Stable Inheritance of IncP Plasmids. In *Promiscuous Plasmids of Gram-Negative Bacteria* (ed. C. M. Thomas), pp. 1–25. London: Academic Press.

Thomas, C. M. & Smith, C. A. (1987). Incompatibility group P plasmids: genetics, evolution and use in genetic manipulation. *Ann. Rev. Microbiol.*, **41**, 77–101.

Thony, B., Hwang, D. S., Fradkin, L. & Kornberg, A. (1991). *iciA*, an *Escherichia coli* gene encoding a specific inhibitor of chromosomal initiation of replication *in vitro*. *Proc. Natl. Acad. Sci.*, **88**, 4066–4070.

Tomizawa, J. (1984). Control of ColE1 plasmid replication: the process of binding RNA I to the primer transcript. *Cell*, **38**, 861–870.

Tomizawa, J. (1986). Control of plasmid ColE1 replication: binding of RNA I to RNA II and inhibition of primer formation. *Cell*, **47**, 89–97.

Tomizawa, J. (1990a). Control of ColE1 plasmid replication. Intermediates in the binding of RNA I and RNA II. *J. Molec. Biol.*, **212**, 683–694.

Tomizawa, J. (1990b). Control of ColE1 plasmid replication. Interaction of Rom protein with an unstable complex formed by RNA I and RNA II. *J. Molec. Biol.*, **212**, 695–708.

Tomizawa, J. & Itoh, T. (1981). Plasmid ColE1 incompatibility determined by interaction of RNA I with primer transcript. *Proc. Natl. Acad. Sci.*, **78**, 6096–6100.

Tomizawa, J. & Som, T. (1984). Control of ColE1 plasmid replication: enhancement of binding of RNA I to the primer transcript by the Rom protein. *Cell*, **38**, 871–878.

Tomizawa, J., Ohmori, H. & Bird, R. E. (1977). Origin of replication of colicin E1 plasmid DNA. *Proc. Natl. Acad. Sci. USA*, **74**, 1865–1869.

Tomizawa, J., Itoh, T., Selzer, G. & Som, T. (1981). Inhibition of ColE1 RNA primer formation by a plasmid-specified small RNA. *Proc. Natl. Acad. Sci. USA*, **78**, 1421–1425.

Travers, A. A. (1993). *DNA–Protein Interactions*. London: Chapman and Hall.

Trawick, J. D. & Kline, B. C. (1985). A two-stage molecular model for control of mini-F replication. *Plasmid*, **13**, 59–69.

Tsuchimoto, S. & Ohtsubo, E. (1989). Effect of the *pem* system on stable maintenance of plasmid R100 in various *Escherichia coli* hosts. *Molec. Gen. Genet.*, **215**, 463–468.

Tsuchimoto, S., Ohtsubo, H. & Ohtsubo, E. (1988). Two genes, *pemK* and *pemI* responsible for stable maintenance of resistance plasmid R100. *J. Bacteriol.*, **170**, 1461–1466.

Tsuchimoto, S., Nishimura, Y. & Ohtsubo, E. (1992). The stable maintenance system *pem* of plasmid R100: degradation of PemI protein may allow PemK protein to inhibit cell growth. *J. Bacteriol.*, **174**, 4205–4211.

Tsutsui, H. & Matsubara, K. (1981) Replication control and switch-off function as observed with a mini-F factor plasmid. *J. Bacteriol.*, **147**, 509–516.

Twigg, A. J. & Sherratt, D. J. (1980). *Trans*-complementable copy-number mutants of plasmid ColE1. *Nature*, **238**, 216–218.

Valone, S. E., Chikami, G. K. & Miller, V. L. (1993). Stress induction of the virulence proteins (SpvA, -B and -C) from native plasmid pSDL2 of *Salmonella typhimurium*. *Infect. Immun.*, **61**, 705–713.

van Biesen, T., Soderbom, F., Wagner, E. G. H. & Frost, L. S. (1993). Structural and functional analyses of the FinP antisense RNA regulatory system of the F conjugative plasmid. *Molec. Microbiol.*, **10**, 35–43.

van Biesen, T. & Frost, L. S. (1994). The FinO protein of IncF plasmids binds FinP antisense RNA and its target, *traJ* mRNA, and promotes duplex formation *Molec. Microbiol.*, **14**, 427–436.

Van Melderen, L., Bernard, P. & Couturier, M. (1994). Lon-dependent proteolysis of CcdA is the key control for activation of CcdB in plasmid-free segregant bacteria. *Molec. Microbiol.*, **11**, 1151–1157.

Vieira, J. & Messing, J. (1982). The pUC plasmids, an M13mp7-derived system for insertion mutagenesis and sequencing with synthetic universal primers. *Gene*, **19**, 259–268.

Vijayakumar, M. N. & Ayalew, S. (1993). Nucleotide sequence analysis of the termini and chromosomal locus involved in site-specific integration of the Streptococcal conjugative transposon Tn*5252*. *J. Bacteriol.*, **175**, 2713–2719.

Wagner, E. G. H., Blomberg, P. & Nordström, K. (1992). Replication control in plasmid R1: duplex formation between the antisenses RNA, CopA, and its target, CopT, is not required for inhibition of RepA synthesis. *EMBO Journal*, **11**, 1195–1203.

Wahle, E. & Kornberg, A. (1988). The partition locus of plasmid pSC101 is a specific binding site for DNA gyrase. *EMBO J.*, **7**, 1889–1895.

Warren, R. L., Womble, D. D., Barton, C. R., Easton, A. M. & Rownd, R. H. (1978). Multiple origins for DNA replication of FII composite R plasmids in *Proteus mirabilis*. In D. D. Schiessinger (eds), *Microbiology 1978* (pp. 96–98). Washington DC: American Society for Microbiology.

Watanabe, E., Inamoto, S., Lee, M.-H., Kim, S. U., Ogua, T., Mori, H., Hiraga, S., Yamasaki, M. & Nagai, K. (1989). Purification and characterization of the *sopB* gene product which is responsible for stable maintenance of mini-F plasmid. *Molec. Gen. Genet.*, **218**, 431–436.

Watanabe, T. (1963). Infective heredity of multiple drug resistance in bacteria. *Bacteriol. Rev.*, **27**, 87–115.

Waters, V. L. & Guiney, D. G. (1993). Processes at the nick region link conjugation, T-DNA transfer and rolling circle replication. *Molec. Microbiol.*, **9**, 1123–1130.

Wilkins, B. M., Rees, C. E., Thomas, A. T. & Read, T. D. (1991). Conjugative events in the recipient cell affecting plasmid promiscuity. *Plasmid*, **25**, 227.

Willetts, N. (1985) Plasmids in *Genetics of bacteria* (eds. J. Scaife, D. Leach, & A. Galizzi), pp. 165–195. London: Academic Press.

Willetts, N. & Skurray, R. (1987). Structure and Function of the F Factor and Mechanism of Conjugation. In *Escherichia coli and Salmonella typhimurium:*

cellular and molecular biology (eds. F. C. Neidhardt), pp. 1110–1133. Washington DC: American Society for Microbiology.

Williams, D. R. & Thomas, C. M. (1992). Active partitioning of bacterial plasmids. *J. Gen. Microbiol.*, **138**, 1–16.

Willshaw, G. A., Smith, H. R., McConnell, M. M. & Rowe, B. (1985). Expression of cloned plasmid regions encoding colonization factor antigen I (CFA/I) in *Escherichia coli. Plasmid*, **13**, 8–16.

Winans, S. C. (1992). Two-way chemical signaling in *Agrobacterium*–plant interactions. *Microbiol. Rev.*, **56**, 12–31.

Witchitz, J. L. & Chabbert, Y. A. (1972). *Ann. Inst. Pasteur, Paris*, **122**, 367–378.

Womble, D. D. & Rownd, R. H. (1986a). Regulation of IncFII plasmid DNA replication. A quantitative model for control of plasmid NR1 replication in the bacterial cell division cycle. *J. Molec. Biol.*, **192**, 529–548.

Womble, D. D. & Rownd, R. H. (1986b). Regulation of lambda-dv plasmid DNA replication. A quantitative model for control of plasmid lambda-dv replication in the bacterial cell division cycle. *J. Molec. Biol.*, **191**, 367–382.

Womble, D. D. & Rownd, R. H. (1987). Regulation of mini-F plasmid DNA replication. A quantitative model for control of plasmid mini-F replication in the bacterial cell division cycle. *J. Molec. Biol.*, **195**, 99–113.

Yamamoto, T., Tanaka, M., Baba, R. & Yamagishi, S. (1981). Physical and functional mapping of Tn2603, a transposon encoding ampicillin, streptomycin, sulfonamide and mercury resistance. *Molec. Gen. Genet.*, **181**, 464–469.

Zaman, S., Radnedge, L., Richards, H. & Ward, J. M. (1993). Analysis of the site for second-strand initiation during replication of the *Streptomyces* plasmid pIJ101. *J. Gen. Microbiol.*, **139**, 669–676.

Zavitz, K. H. & Marians, K. J. (1991). Dissecting the functional role of PriA protein-catalysed primosome assembly in *Escherichia coli* DNA replication. *Molec. Microbiol.*, **5**, 2869–2873.

Zinder, N. D. & Lederberg, J. (1952). Genetic exchange in *Salmonella. J. Bacteriol.*, **64**, 679–699.

Zund, P. & Lebek, G. (1980). Generation time-prolonging R plasmids: correlation between increases in the generation time of *Escherichia coli* caused by R plasmids and their molecular size. *Plasmid*, **3**, 65–69.

Zyskind, J. W. & Smith, D. W. (1986). The bacterial origin of replication, *oriC. Cell*, **46**, 489–490.

Index